AF872661

Electrochemistry and Electrolysis for Gold and Silver Recovery for Chemists and Precious Metal Refiners

Roggie Clark

Continue Learning At Our Website: Roggieclarkbooks.com

Continue exploring educational resources, tools, books, workbooks, and learning materials designed to help readers continue building knowledge and practical skills. Visit: www.roggieclarkbooks.com

Inside you'll find:

- Educational worksheets
- Beginner-friendly guides
- Printable activities
- Companion materials for books and courses

A growing library of educational content for readers, collectors, learners, parents, teachers, and lifelong students.

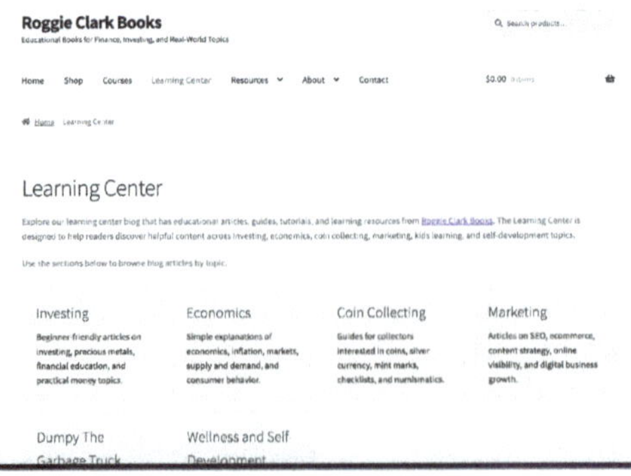

Copyright

Title: Electrochemistry and Electrolysis for Gold and Silver Recovery for Chemists and Precious Metal Refiners
Author: Roggie Clark

Copyright © 2026 by Roggie Mullis Clark III

All rights reserved. No part of this book may be reproduced, stored in a retrieval system, or transmitted in any form or by any means, electronic, mechanical, photocopying, recording, scanning, or otherwise, without the prior written permission of the author and publisher except in the case of brief quotations embodied in critical reviews and certain other noncommercial uses permitted by copyright law.

All trademarks, trade names, or logos mentioned or used in this book are the property of their respective owners and are used for identification purposes only.

Cover Design: Roggie Clark
Editing: Renee' Clark
Printed in the United States of America

Disclaimer

This book is provided for **educational and informational purposes only**. It is intended to explain electrochemistry, electrolysis, precious metal recovery, and related refining concepts for chemists, chemistry learners, educators, precious metal refiners, and other technically interested readers.

The information presented is not a substitute for professional laboratory training, chemical safety training, engineering advice, environmental guidance, or instruction from a qualified chemist, metallurgist, industrial hygienist, or precious metal refining professional.

Electrochemistry and precious metal recovery can involve significant hazards. Depending on the process, these may include **corrosive acids and bases, reactive chemicals, toxic or irritating gases, dissolved heavy metals, electrical equipment, hot solutions, pressurized or sealed systems, high-temperature equipment, open flames, and molten metals**. Chemical reactions can also behave differently when concentrations, temperatures, materials, equipment, or operating conditions are changed.

Procedures described in this book should not be interpreted as a guarantee that a particular recovery or refining process is safe, appropriate, profitable, or suitable for a specific material or environment. Readers are responsible for evaluating their equipment, chemicals, materials, experience, facilities, and applicable laws before performing any practical work.

Never assume that an unknown metal, solution, powder, sediment, residue, electronic component, or scrap material is safe to process. Material composition should be properly identified before chemical or electrochemical treatment.

Some processes discussed in this book may require professional laboratory facilities, engineering controls, appropriate ventilation or fume extraction, specialized personal protective equipment, chemical-resistant equipment, and regulated waste disposal. Processes that cannot be performed safely with the available facilities and training should not be attempted.

Gold- and silver-bearing scrap can contain other metals and substances that introduce additional hazards. Electronic and industrial materials may contain **lead, nickel, cadmium, chromium compounds, copper, tin, and other potentially hazardous substances**. Heating, dissolving, grinding, melting, or electrochemically processing unknown materials can create additional exposure risks.

The author and publisher make no warranties regarding the accuracy, completeness, recovery yield, purity, profitability, or suitability of any process for a particular application. Neither the author nor the publisher assumes responsibility for injury, illness, property damage, environmental contamination, financial loss, equipment damage, regulatory violations, or other consequences arising from the use or misuse of the information presented in this book.

Always follow applicable local, state, and federal regulations concerning chemical handling, workplace safety, electronic waste, hazardous waste, wastewater, air emissions, precious metal processing, and environmental protection.

Safety Precautions

Electrochemical precious metal recovery should be approached as **laboratory and refining work**, not as an ordinary household activity. Understanding the chemistry of a process is only one part of performing it responsibly. Readers must also understand the hazards created by the chemicals, electricity, equipment, metals, and reaction products involved.

Chemical Safety

Before using any chemical, obtain and read its current **Safety Data Sheet (SDS)**. Pay particular attention to incompatibilities, required personal protective equipment, exposure limits, first-aid information, storage requirements, and disposal instructions.

Never mix chemicals simply because a recovery demonstration or procedure appears to use them together. Changing concentration, order of addition, temperature, contamination, or quantity can substantially change a reaction.

Concentrated acids and bases can cause severe chemical burns and eye damage. Some chemical combinations can release dangerous gases or generate substantial heat.

Use properly rated chemical-resistant containers and equipment. Household food containers, drinking glasses, cookware, and food preparation equipment should not be used for precious metal refining.

Never eat, drink, smoke, or prepare food in an area used for chemical processing.

Ventilation and Gas Hazards

Electrolysis can produce gases at electrode surfaces. The gases produced depend on the **electrolyte, electrode materials, voltage, current, contaminants, and chemical species present**.

Do not assume that bubbles are harmless hydrogen or oxygen.

For example, chloride-containing electrolytes can create conditions in which chlorine may be generated at an electrode. Acidic systems and other electrolyte compositions can introduce additional hazards.

Processes capable of producing hazardous fumes or gases require appropriate professional ventilation or fume-control equipment. An open window, household fan, or ordinary room ventilation should not be assumed sufficient.

Never intentionally smell a reaction vessel to identify a gas.

Electrical Safety

Electrolysis combines liquids with electrical equipment, creating hazards that require careful equipment selection and operating practices.

Use power supplies appropriate for electrochemical work and keep electrical connections protected from spills and accidental contact.

Turn off and disconnect electrical power before adjusting electrodes, removing materials, repositioning conductive components, or servicing the cell.

Electrodes should not be allowed to contact one another accidentally while the system is energized.

Wet hands, damaged wires, improvised electrical connections, and inappropriate power supplies increase risk and should be avoided.

Personal Protective Equipment

Appropriate **personal protective equipment (PPE)** depends on the chemicals and operations involved and should be selected according to the applicable SDS information and professional safety guidance.

Depending on the operation, this may include chemical splash goggles, appropriate gloves, protective clothing, closed footwear, face protection, and additional specialized equipment.

PPE is not a replacement for ventilation or other engineering controls.

Heavy Metals and Contaminated Materials

Precious metal-bearing materials frequently contain substances other than gold and silver.

Electronic scrap, electrical contacts, alloys, industrial residues, and plated materials can contain potentially hazardous metals and compounds. Dust, solutions, filters, residues, rinse water, and recovered solids should therefore be treated as potentially contaminated until their composition is understood.

Avoid creating unnecessary dust.

Do not dispose of metal-bearing solutions, acids, bases, rinses, sludge, or refining residues in sinks, toilets, storm drains, soil, or ordinary household waste unless specifically permitted by applicable regulations.

Heating and Melting

Heating chemical solutions can accelerate reactions and increase vapor or gas production. A reaction that appears manageable at room temperature can behave differently when heated.

Molten gold, silver, copper, and other metals present severe burn and fire hazards. Moisture introduced into high-temperature operations can also create dangerous conditions.

Metal melting should be performed only with equipment, protective systems, facilities, and training appropriate for high-temperature metallurgical work.

Never Process Unknown Materials Blindly

Before processing a material, determine as much as reasonably possible about its composition.

Do not assume a silver-colored surface contains silver or that a gold-colored surface contains gold.

Likewise, do not assume that a recovered black, gray, brown, or metallic powder consists entirely of the desired precious metal.

Unknown feed materials can introduce unexpected reactions and hazardous contaminants.

Separate Learning From Experimentation

Chemistry students and educators can use the recovery processes discussed in this book as **case studies for electrochemistry** without physically reproducing every procedure.

Electrode reactions can be analyzed, half-reactions balanced, theoretical yields calculated, process variables compared, and recovery systems evaluated without performing hazardous refining operations.

Classroom demonstrations should be selected according to institutional laboratory policies, available safety equipment, instructor qualifications, student experience, and applicable chemical safety requirements.

Prepare Before Beginning

Before performing practical electrochemical or refining work, know:

- What materials and chemicals are present
- What reactions are expected
- What competing reactions are possible
- What gases or byproducts could be generated
- What PPE and engineering controls are required

- How spills and exposures will be handled
- How the process can be stopped safely
- How all resulting solutions, solids, filters, and waste will be stored and disposed of

If these questions cannot be answered confidently, the process should not begin.

Precious metal recovery is never worth risking serious injury, toxic exposure, fire, environmental contamination, or unsafe chemical handling. Technical knowledge should always be accompanied by appropriate training, equipment, facilities, and responsible laboratory practices.

Image Credits and Public Domain Notice

All images, illustrations, graphics, photographs, visual references, and related materials included in this publication are believed to be in the public domain, used under applicable licenses or permissions, created by the author or publisher, or obtained from sources that clearly identify them as available for lawful use.

The author and publisher have made a good faith effort to use only visual materials that are lawfully available for use and distribution. Where applicable, images and visual materials may have been resized, cropped, edited, enhanced, stylized, modified, reformatted, or arranged for educational, informational, design, formatting, or presentation purposes.

This publication is intended for educational, informational, and illustrative purposes only. The inclusion of any image, artwork, photograph, visual reference, symbol, historical material, design, or graphic element is not intended to claim ownership over public domain materials or any third-party intellectual property rights that remain with their respective owners.

If you believe any image or visual material included in this publication has been used in error, please contact the author or publisher so the matter may be reviewed and corrected in future editions or releases if necessary.

Any factual information, specifications, historical references, values, descriptions, statistics, or related content accompanying visual materials are provided for general informational purposes only and may change over time. Readers should conduct their own independent research and consult appropriate sources or qualified professionals when making decisions based on information presented in this publication.

This publication is independently produced and is not affiliated with, endorsed by, sponsored by, or associated with any company, organization, government agency, institution, trademark holder, or third party unless specifically stated otherwise.

Table of Contents

Introduction: Electrochemistry, Electrolysis, and Precious Metal Recovery

Gold and silver recovery offers a particularly useful way to see electrochemistry move beyond equations and diagrams and into a real material process. When electricity is introduced into a properly designed chemical system, metals can behave in remarkably different ways. One metal may dissolve from an electrode, another may remain behind, and dissolved metal ions may be converted back into solid metal. These processes demonstrate why **electrochemistry** is important not only to chemists, but also to precious metal refiners who need to understand what is occurring inside a recovery system. This book approaches gold and silver recovery from both perspectives. The goal is to explain the chemistry clearly while following the practical sequence of preparing, separating, recovering, purifying, and refining precious metals through electrochemical processes.

Connecting Electrochemical Science to Metal Recovery

At its foundation, electrochemistry examines the relationship between chemical reactions and electrical energy. Of particular importance to precious metal recovery are reactions involving the transfer of electrons between chemical species. These reactions can change the chemical form of a metal and, under the right conditions, make it possible to separate one material from another.

Electrolysis is especially important because it uses an external electrical source to drive chemical changes. Instead of simply waiting for a reaction to occur spontaneously, an electrolytic system supplies electrical energy that can promote oxidation and reduction at different locations within the cell.

For precious metal refiners, this creates possibilities that mechanical separation alone cannot provide. A piece of material may contain copper, silver, gold, and other substances so closely associated that simply cutting, grinding, or sorting the material will not produce a clean separation. Electrochemical processing approaches the problem according to the chemical behavior of the metals.

An electrolytic system generally contains an electrolyte, electrodes, electrical connections, and a direct-current power source. Once the system is energized, reactions occur at the electrode surfaces.

This is where terminology becomes important.

Oxidation occurs at the anode and involves the loss of electrons. **Reduction** occurs at the cathode and involves the gain of electrons. In an electrolytic cell, the anode is connected to the positive terminal of the external power supply and the cathode to the negative terminal.

The electrolyte provides a medium through which ions can move. The electrical circuit therefore involves electron movement through the external conductors and ionic movement through the electrolyte.

These principles are straightforward when written as definitions. Their significance becomes much clearer when they are connected to actual metals.

A metal atom can lose electrons and become an ion:

$$M \rightarrow M^{n+} + ne^-$$

A dissolved metal ion can gain electrons and return to its metallic state:

$$M^{n+} + ne^- \rightarrow M$$

Those two generalized reactions represent much of what makes electrochemical metal processing possible. The challenge is controlling the system so that the reactions occurring are the ones that contribute to the desired separation or recovery.

Electrolytic Cell

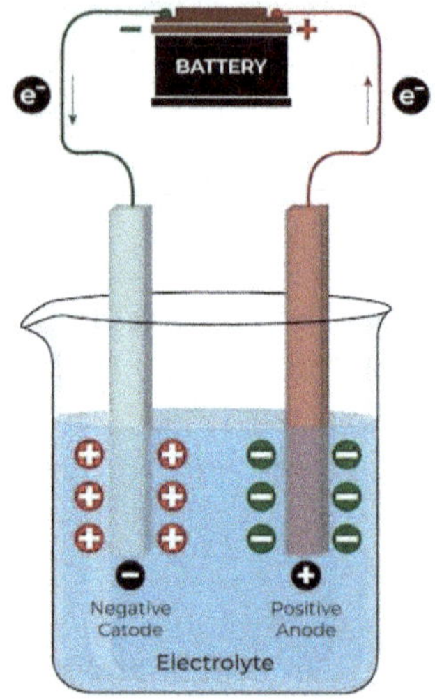

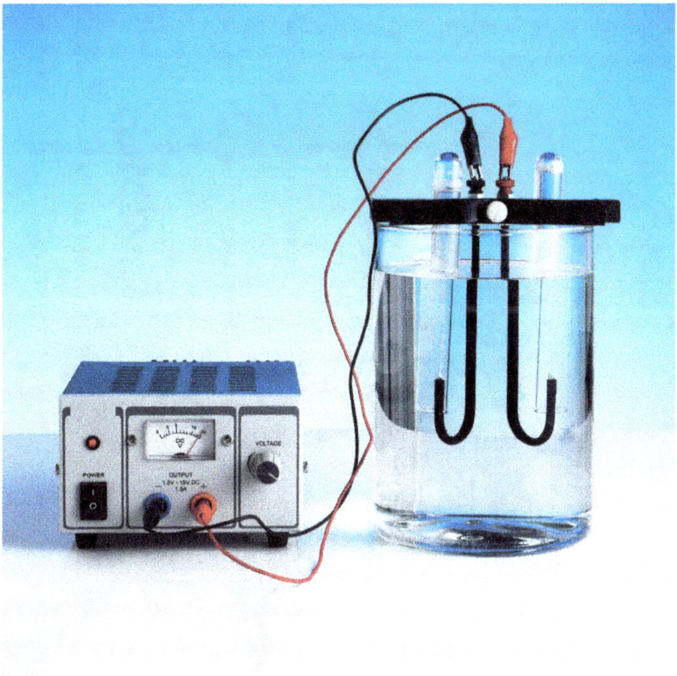
DC
V
POWER
VOLTAGE
OUTPUT

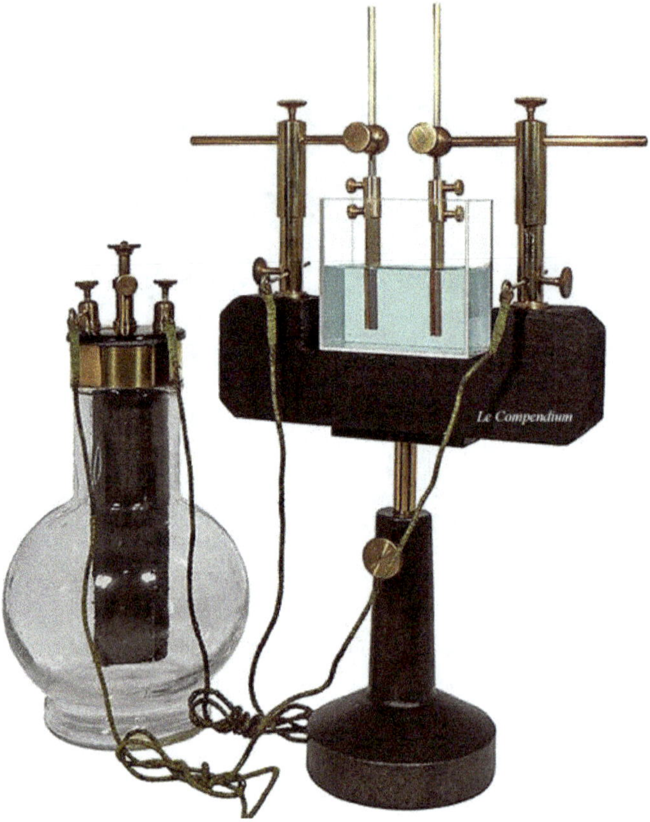
Le Compendium

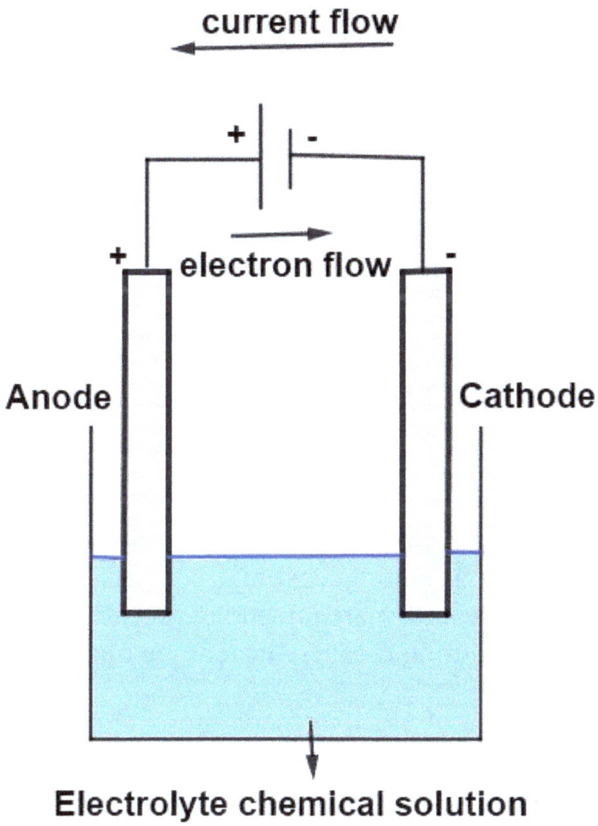

For chemists and chemistry learners, precious metal recovery therefore provides a physical context for concepts that can otherwise seem abstract. For refiners, electrochemistry provides an explanation for changes they may observe during processing. The two perspectives meet inside the electrolytic cell.

Why Gold and Silver Are Useful Electrochemical Case Studies

Gold and silver are particularly interesting because their value makes efficient separation and recovery important, while their chemistry provides useful examples of how electrochemical behavior can be controlled.

They are also frequently encountered alongside other metals.

Silver may exist as plating or as part of electrical contacts and other metal-bearing components. Gold can appear as plating, coatings, contacts, jewelry alloys, industrial material, and in prepared solutions produced during recovery or refining operations.

The amount of precious metal present may be small compared with the amount of base metal surrounding it. That creates a separation problem.

Consider silver plating on a copper substrate. Visually, the valuable material appears on the surface, but mechanically removing every trace of silver can be difficult and inefficient. Electrochemical processing offers another way of approaching the material by exploiting differences in how the metals behave under controlled conditions.

The silver recovery example examined later in this book begins with silver-bearing copper material and uses electrical current as part of the separation process. The visible result is the exposure of the underlying copper and the collection of silver-bearing material that can then undergo additional purification.

Gold provides a different electrochemical case study.

Rather than assuming every gold-bearing object can simply be placed into an electrolytic cell, it is important to distinguish between **recovery from a prepared gold-bearing system** and the earlier chemical or mechanical operations that produced that system. Electrolysis can then become one stage within a larger recovery sequence.

This distinction is central to the approach taken throughout this book.

Recovery is not necessarily the same as **refining**.

Recovery concentrates or captures the desired metal from the material in which it was previously contained. Refining seeks to increase the purity of that recovered metal by separating it from remaining contaminants.

An electrochemical operation can contribute to either objective depending on how the system is designed.

This is why simply observing a dark powder, metallic deposit, sediment, or changing electrode does not establish that a pure precious metal has been produced. The material must be understood within the chemistry of the process. Other metals may be present, competing reactions may occur, and additional purification may be necessary.

Gold and silver therefore allow us to explore a larger question:

How can electrical and chemical conditions be controlled so that a complicated metal-bearing material becomes a more concentrated and eventually more purified precious metal product?

That question will guide the practical chapters that follow.

Understanding the Recovery and Refining Workflow

Precious metal recovery should be understood as a **workflow**, rather than as a single reaction.

The starting material comes first. Its composition influences nearly every decision that follows. Silver-plated copper, for example, presents a different problem from a gold-bearing solution. The appropriate electrolyte, electrode configuration, operating conditions, collection method, and purification sequence depend on what is being processed.

The next stage is preparation.

Material may need to be sorted, cleaned, separated from unwanted components, or otherwise prepared before electrolysis is appropriate. A recovery system cannot compensate for poor understanding of the feed material. Introducing unknown or incompatible materials can create contamination, unwanted reactions, and serious safety problems.

The electrolytic cell must then be configured for the intended reaction.

Electrode material, electrode placement, electrolyte composition, voltage, current, temperature, surface area, and processing time can all influence what occurs. More electrical power does not automatically produce better recovery. Increasing voltage or current can change reaction rates, promote competing reactions, generate excessive gas, increase heating, and reduce selectivity.

The electrolysis stage therefore requires observation and control.

Changes at the electrodes matter. Changes in solution matter. Gas evolution matters. Deposits, sediment, discoloration, exposed base metal, and changes in electrical behavior can all provide information about what is occurring.

But electrolysis is not necessarily the end of the process.

Recovered material may contain the desired precious metal together with copper, oxides, salts, electrolyte residues, or other contaminants. Collection is therefore followed by an evaluation of what has actually been recovered.

Depending on the system, additional separation and purification may be required before the material can reasonably be considered refined gold or silver.

This distinction becomes particularly clear in silver recovery. Electrolysis can help separate silver-bearing material from its substrate, but the collected product may still require further processing to isolate and purify the silver. The complete process therefore moves through several stages rather than ending when material first appears at the bottom of a vessel.

The same principle applies to gold. A successful electrochemical reaction must be followed by appropriate collection, washing, purification, verification, and, where applicable, final refining.

These later operations also introduce hazards beyond the electrical system itself. Precious metal processing can involve corrosive chemicals, toxic or irritating fumes, reactive solutions, electrical equipment, high temperatures, and molten metal. Advanced knowledge does not remove those hazards. Appropriate laboratory controls, ventilation, personal protective equipment, chemical compatibility knowledge, waste management, and applicable environmental and workplace requirements remain essential.

Throughout the chapters ahead, the process will therefore be examined from two directions simultaneously.

We will ask **what is being done**, because the book is intended to follow the recovery process step by step.

We will also ask **why it works**, because understanding the underlying electrochemistry makes those steps meaningful.

Last Thoughts

Electrolysis provides a bridge between electrochemical theory and the physical recovery of valuable metals. Gold and silver make that connection especially visible because their separation, recovery, purification, and refining require careful attention to both chemistry and process control. The chapters ahead will begin with the electrochemical principles needed to understand these systems before moving into equipment, silver recovery and purification, gold recovery, and finally the variables that determine efficiency and purity. The objective is not simply to reproduce a sequence of steps, but to understand the reactions well enough to recognize **what is happening, why it is**

happening, and how each stage contributes to the larger recovery and refining process.

Chapter 1: Electrochemistry and Electrolysis Fundamentals

Precious metal recovery through electrolysis depends on controlling chemical reactions with electrical energy. Before examining the equipment or following the silver and gold recovery processes, it is important to understand what is occurring inside the electrolytic cell. The movement of electrons, the behavior of ions, and the reactions occurring at electrode surfaces determine whether a metal dissolves, remains in place, deposits onto another surface, or participates in an unwanted competing reaction. These fundamentals provide the framework for understanding every recovery process discussed later in the book.

Oxidation, Reduction, and Electron Transfer

Electrochemical reactions involve the transfer of electrons. Although a recovery system may contain several metals and dissolved chemical species, the underlying reactions can be understood by following where electrons are lost and where they are gained.

As introduced earlier, oxidation involves the loss of electrons and reduction involves their gain. These two processes occur together. Electrons released through one reaction must ultimately participate in another reaction elsewhere in the electrochemical system.

For a generalized metal, oxidation can be represented as:

$$M \rightarrow M^{n+} + ne^-$$

A neutral metal atom loses electrons and becomes a positively charged ion.

The reverse process can be represented as:

$$M^{n+} + ne^- \rightarrow M$$

Here, a dissolved metal ion gains electrons and becomes neutral metallic material.

These equations are simple, but they describe an important mechanism behind metal dissolution and deposition.

Consider copper as an example. Copper metal can undergo oxidation according to:

$$Cu \rightarrow Cu^{2+} + 2e^-$$

Copper atoms leave the metallic structure as copper ions while releasing electrons. Under appropriate conditions, copper ions can undergo the reverse reaction:

$Cu^{2+} + 2e^- \rightarrow Cu$

The ions gain electrons and become metallic copper.

Silver can similarly participate in an oxidation-reduction pair:

$Ag \rightarrow Ag^+ + e^-$

and:

$Ag^+ + e^- \rightarrow Ag$

The exact reactions occurring in a real recovery system depend on the chemical environment. The presence of complexing agents, acids, salts, other metals, and competing ions can change which species exist in solution and how readily a particular reaction proceeds.

This is one reason precious metal recovery cannot be understood simply by remembering that electricity "pulls metal out" of a solution.

Electricity supplies the driving force, but the chemistry determines what reactions are available.

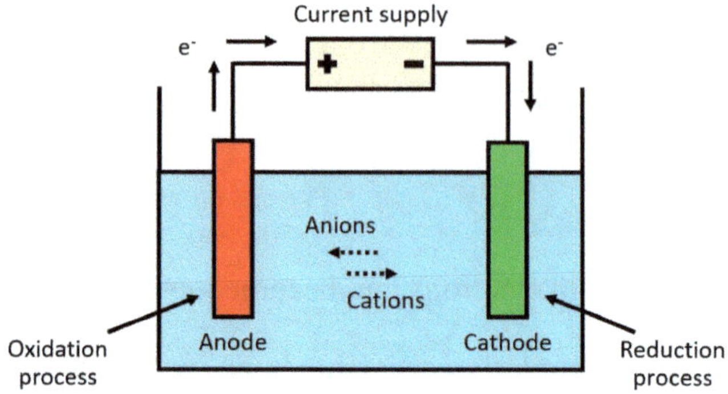
Current supply
e⁻
e⁻
Anions
Cations
Oxidation process
Anode
Cathode
Reduction process

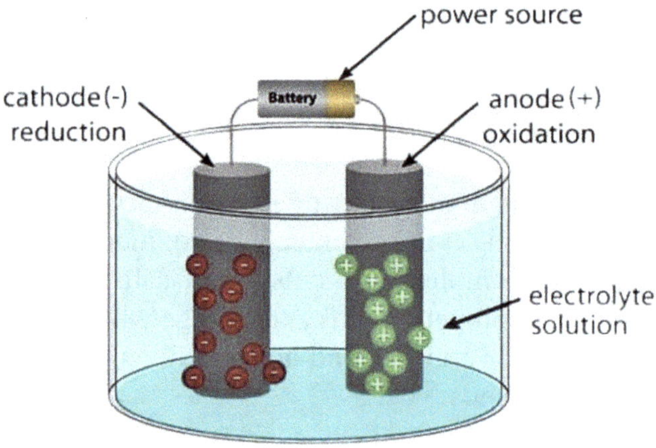
power source
Battery
cathode(-)
reduction
anode(+)
oxidation
electrolyte
solution

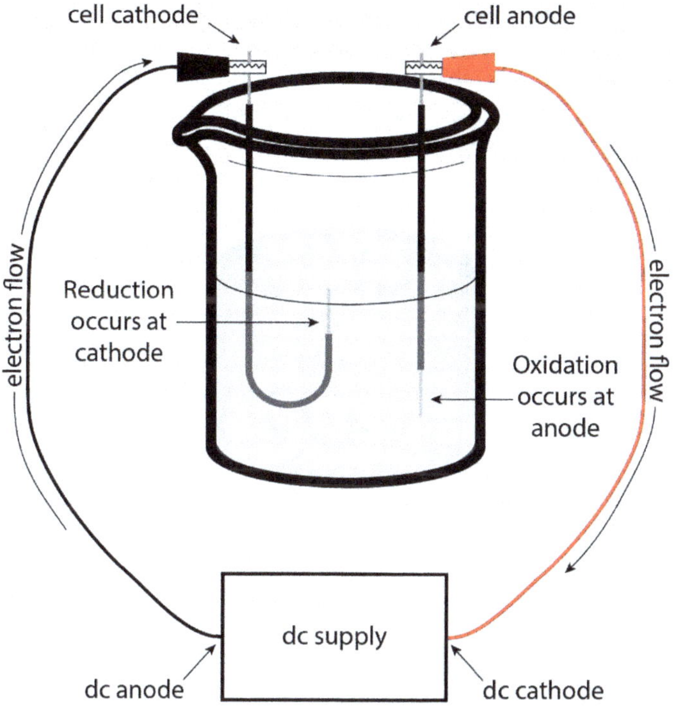
cell cathode
cell anode
electron flow
Reduction occurs at cathode
Oxidation occurs at anode
electron flow
dc supply
dc anode
dc cathode

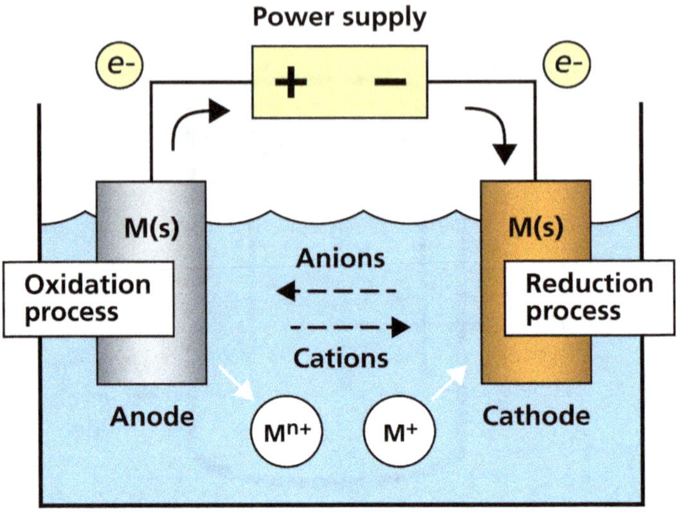

A useful way to think about an electrolytic recovery system is as a controlled exchange.

At one location, electrons are being removed through oxidation. At another location, electrons are being supplied for reduction. Meanwhile, ions move through the electrolyte to maintain the electrochemical processes occurring inside the cell.

For precious metal recovery, the objective is to make those reactions useful.

A refiner may want a particular metal to enter solution while another remains relatively unaffected. In another system, the goal may be to reduce dissolved precious metal species so that metallic material can be collected.

This introduces the idea of **selectivity**.

An electrochemical system is most useful when its operating conditions favor the desired reaction over competing reactions. If several metals can react under the conditions being used, the resulting recovered material may contain significant contamination.

Gold, silver, copper, nickel, and other metals do not behave identically. Their different electrochemical properties are part of what makes separation possible, but those differences must be considered together with the actual chemical forms present in the system.

For chemistry learners, this is an important transition from textbook redox equations to real electrochemistry. A balanced half-reaction tells us what electron transfer is possible. A working recovery process must also account for the environment in which that reaction occurs.

For refiners, the same principle explains why visually similar starting materials may behave differently. Their underlying compositions may not be the same.

Understanding electron transfer therefore gives us the first piece of the recovery process. The next question is where these reactions actually occur.

Anodes, Cathodes, and Electrolytes

The reactions of an electrolytic cell occur at interfaces between the electrodes and the electrolyte.

The anode is the location of oxidation.

The cathode is the location of reduction.

In an electrolytic cell powered by an external direct-current source, the anode is connected to the positive terminal and the cathode is connected to the negative terminal.

That distinction matters enormously in metal recovery because reversing the intended electrode arrangement changes the electrochemical conditions at the material surfaces.

The electrodes themselves can also participate differently depending on the system.

An electrode that participates chemically in the reaction may dissolve, change composition, or acquire a deposit. An electrode selected primarily to provide a conductive surface may be intended to remain comparatively unchanged, although no electrode material should automatically be assumed inert under every chemical condition.

Between the electrodes is the electrolyte.

An electrolyte contains mobile ions capable of carrying charge through the liquid portion of the cell. Without sufficient ionic conductivity, the electrochemical circuit cannot operate effectively.

This creates two interconnected pathways.

Electrons travel through the external electrical circuit.

Ions move through the electrolyte.

Both are necessary for sustained electrochemical operation.

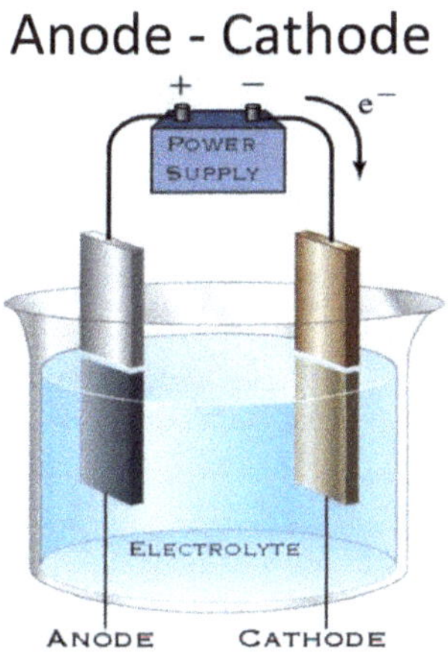

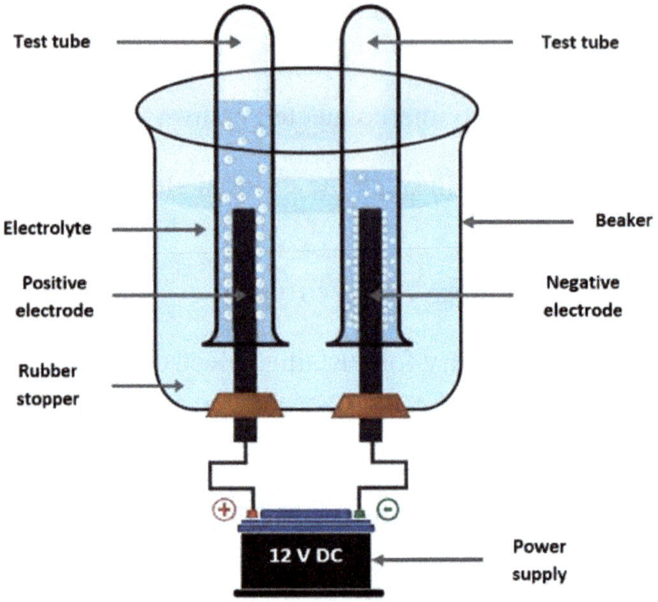
Test tube
Test tube
Electrolyte
Beaker
Positive
electrode
Negative
electrode
Rubber
stopper
12 V DC
Power
supply

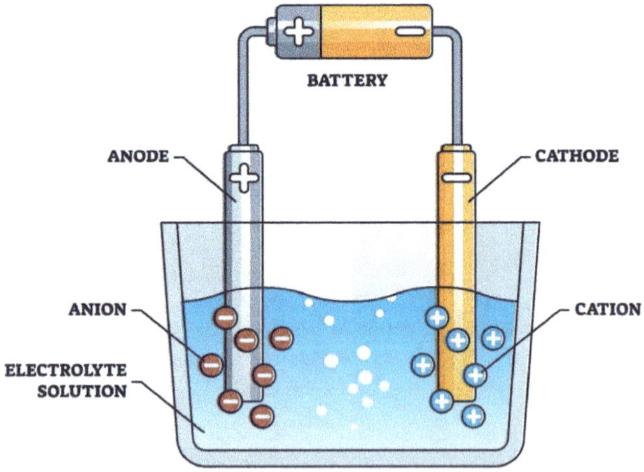
BATTERY
ANODE
CATHODE
ANION
CATION
ELECTROLYTE
SOLUTION

Electrolytic Cell

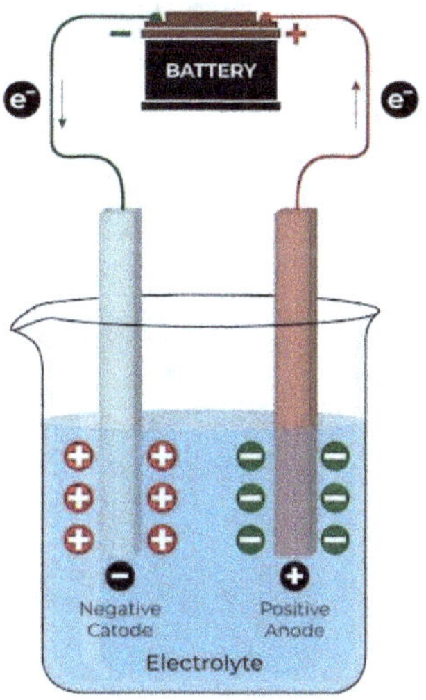

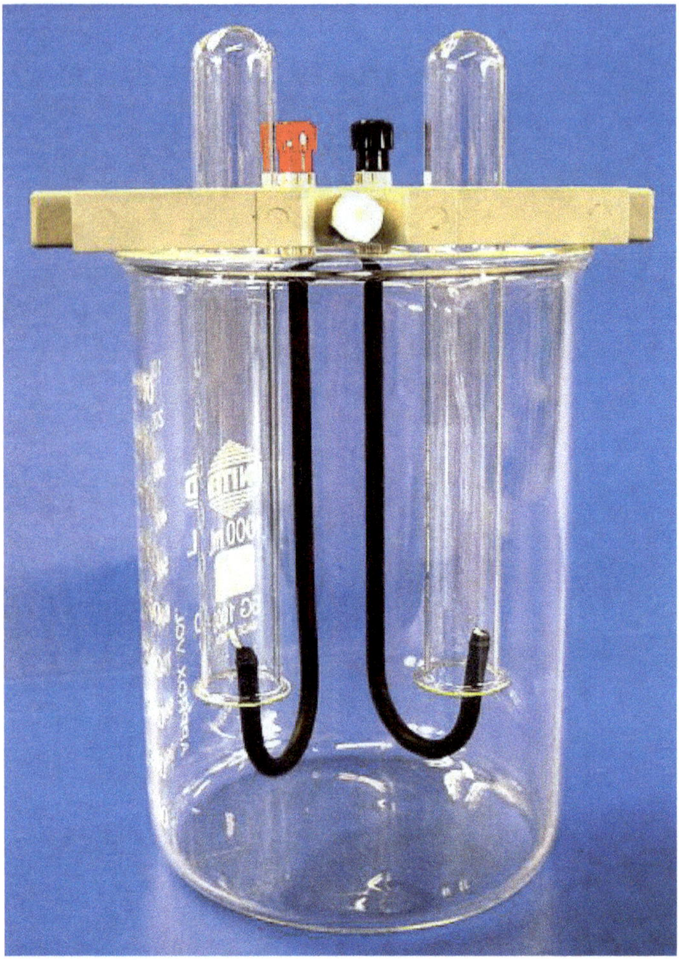

Imagine two electrodes submerged in an electrolyte and connected to a DC power supply.

At the cathode, electrons supplied through the external circuit become available for reduction reactions. If an appropriate dissolved metal species reaches the cathode surface and the conditions favor its reduction, it may gain electrons.

At the anode, oxidation reactions release electrons into the external circuit.

However, metal deposition is not the only reaction that can occur.

Water and other chemical species may also undergo electrochemical reactions. Gas may evolve. An electrode may corrode. A dissolved species other than the desired precious metal may be reduced. Multiple reactions may even occur simultaneously.

This is why bubbles should not automatically be interpreted as evidence that precious metal recovery is proceeding efficiently.

Gas formation can indicate a competing electrochemical reaction.

Likewise, a deposit on an electrode should not automatically be identified as pure gold or silver simply because those metals were present somewhere in the starting material.

The identity of a deposit depends on the ions available, their concentrations, the electrode conditions, the electrolyte chemistry, and the electrical conditions applied to the cell.

The **electrode surface** itself is therefore an important reaction zone.

Its area affects how the applied current is distributed. Its composition can influence reactions. Deposits can change its surface over time. Contamination can interfere with the intended process.

Electrode spacing also matters. The physical geometry of the cell influences electrical resistance and current distribution, while excessive proximity can introduce practical problems.

These factors help explain why an electrolytic recovery system should be viewed as more than a container with two pieces of metal connected to a power supply.

It is a controlled electrochemical environment.

Understanding that environment becomes especially important when moving from a small demonstration to a larger recovery operation. A configuration that appears to work with a small amount of material may behave differently when electrode area, electrolyte volume, metal concentration, or processing load changes.

The underlying electrochemistry remains the same, but the operating conditions have changed.

Voltage, Current, and Controlling Electrochemical Reactions

A power supply provides the electrical energy needed to drive an electrolytic process, but voltage and current describe different aspects of that process.

Voltage represents electrical potential difference. In an electrolytic system, sufficient applied potential is required to drive the intended nonspontaneous reactions while also overcoming practical losses within the cell.

Current represents the rate of electrical charge flow.

The two are related, but they should not be treated as interchangeable controls.

Applying more voltage can increase current under a given set of conditions, but that does not mean continuously increasing voltage improves precious metal recovery.

Once the applied potential becomes sufficiently large, additional reactions may become favorable. Water may undergo electrolysis more aggressively, gases may evolve faster, electrodes may behave differently, and unwanted metals or other species may begin participating in the process.

This is one reason an electrolytic cell can change dramatically when its electrical conditions are altered.

The reference silver-recovery process demonstrates this visually. The operator initially works at a lower voltage and observes material separating from the plated copper. When the voltage is increased, substantially greater bubbling and current are observed. The example is useful because it demonstrates that increasing electrical input changes more than processing speed.

A more aggressive cell is not necessarily a more selective cell.

Another important measurement is **current density**, which relates current to electrode surface area.

It can be expressed generally as:

Current Density = Current / Electrode Area

This provides more information than current alone.

For example, five amperes distributed across a large electrode surface does not create the same local electrochemical conditions as five amperes concentrated onto a much smaller surface.

Current density can influence the structure and quality of a metal deposit, reaction rates, gas evolution, heating, and the likelihood of competing reactions.

Time introduces another important relationship.

According to **Faraday's laws of electrolysis**, the theoretical quantity of material transformed at an electrode is related to the amount of electrical charge passed through the system.

Electrical charge can be expressed as:

$Q = I \times t$

where:

Q = charge

I = current

t = time

This relationship is extremely useful because it connects electrical measurements to chemical change.

In an idealized system, the amount of metal deposited or dissolved can therefore be predicted from the charge transferred and the number of electrons involved in the relevant reaction.

Real recovery systems are less perfect.

Some of the electrical current may support unwanted reactions instead of the desired metal reaction. Gas evolution, competing metal deposition, impurities, changing concentrations, electrode condition, and other factors can reduce **current efficiency**.

That difference between theoretical and actual recovery is important.

If a calculation predicts that a certain amount of metal should have been deposited but substantially less is collected, the discrepancy tells us something about the process. Current may have been consumed by competing reactions, the starting assumptions may have been incorrect, or some recovered material may have been lost during collection and purification.

Electrical measurements therefore provide more than operating numbers. They can help diagnose the electrochemical system.

Temperature should also be considered.

Electrical resistance and chemical reactions can produce heat, while temperature itself can influence conductivity, reaction rates, diffusion, and other chemical behavior. An unexpected rise in temperature can therefore indicate changing cell conditions.

This leads to one of the most important principles in electrochemical precious metal recovery:

Control is more valuable than maximum power.

The goal is not simply to make the reaction occur as rapidly as possible. The goal is to create conditions that favor the desired electrochemical pathway while limiting contamination, competing reactions, excessive heating, and unnecessary energy use.

This principle will become increasingly important when we examine actual silver and gold recovery systems. The power supply is not merely an on-and-off device. It is one of the tools used to control what happens inside the cell.

Last Thoughts

Electrochemical precious metal recovery begins with a relatively small group of principles: electrons are transferred through oxidation and reduction, those reactions occur at the anode and cathode, ions move through the electrolyte, and an external power supply provides the electrical conditions needed to drive electrolysis. The complexity appears when these principles are applied to real materials containing multiple metals and competing chemical species. Voltage, current, electrode area, electrolyte composition, time, and other operating conditions can determine whether the desired reaction dominates or becomes only one of several reactions occurring simultaneously. With these fundamentals established, the next chapter moves from the chemistry itself to **building and understanding the electrolytic recovery system that puts these principles into practice**.

Chapter 2: Building and Understanding an Electrolytic Recovery System

An electrolytic recovery system turns the principles from the previous chapter into a working process. Although the basic arrangement may appear simple, two electrodes, an electrolyte, a container, and a power source, each component affects what happens inside the cell. Precious metal recovery adds another layer of complexity because the starting material may contain several metals, coatings, alloys, and contaminants. A well-understood system therefore begins before electrical power is applied. The material must be evaluated, the cell must be configured for the intended reaction, and the process must be monitored closely enough to recognize both useful and unwanted changes.

Electrolytic Cells, Electrodes, and Power Supplies

An **electrolytic cell** provides the physical environment in which externally supplied electrical energy drives the desired chemical reactions. Its design does not have to be complicated, but its components must be chemically and electrically appropriate for the process being studied.

At the center of the system are the electrodes.

One electrode provides the location for oxidation, while the other provides the location for reduction. Depending on the recovery method, an electrode may consist of the metal-bearing material itself, a conductive support, or another material selected for its behavior in the electrolyte.

The choice of electrode material matters because an electrode should never be considered chemically neutral simply because it conducts electricity.

Stainless steel, graphite, copper, and various other conductive materials can behave differently depending on electrolyte composition, applied potential, temperature, and the substances present in the cell. An electrode suitable for one system may contaminate or interfere with another.

The reference silver process, for example, uses silver-plated copper material together with stainless steel electrodes. The plated material is bundled with copper wire to provide electrical contact before being submerged in the electrolyte.

This highlights an easily overlooked requirement: **electrical contact**.

If several pieces of scrap are being processed together, simply placing them near one another does not guarantee that they are electrically connected. Poor contact can cause some surfaces to participate strongly while other pieces remain largely inactive.

The physical arrangement of the material therefore matters.

Electrodes should also be positioned so that they do not accidentally touch one another. Direct contact can create a short circuit that bypasses the intended electrochemical pathway.

The container completes the physical arrangement.

For laboratory-scale observation, transparent vessels can make changes easier to see, but vessel material must be compatible with the electrolyte and operating conditions. A container should not be selected merely because it is convenient.

Above all, chemical compatibility should guide equipment selection.

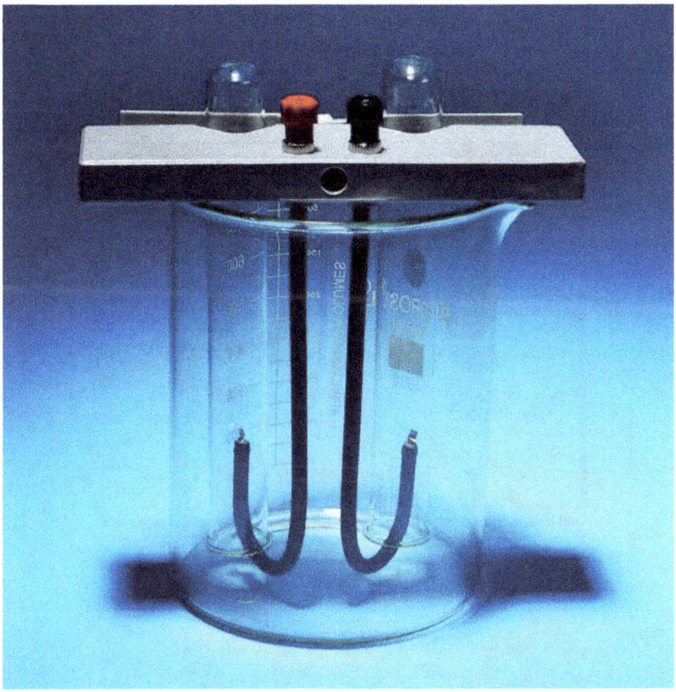

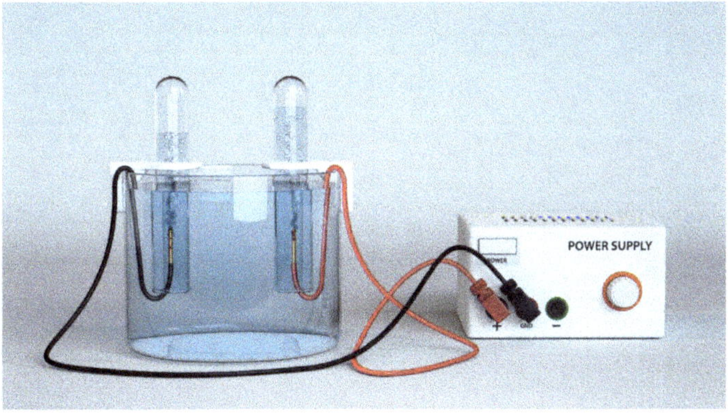
POWER SUPPLY

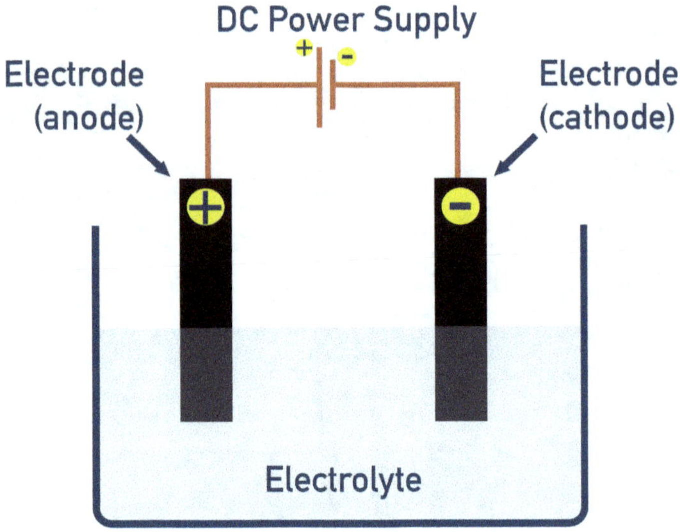
DC Power Supply
+
-
Electrode
(anode)
Electrode
(cathode)
Electrolyte

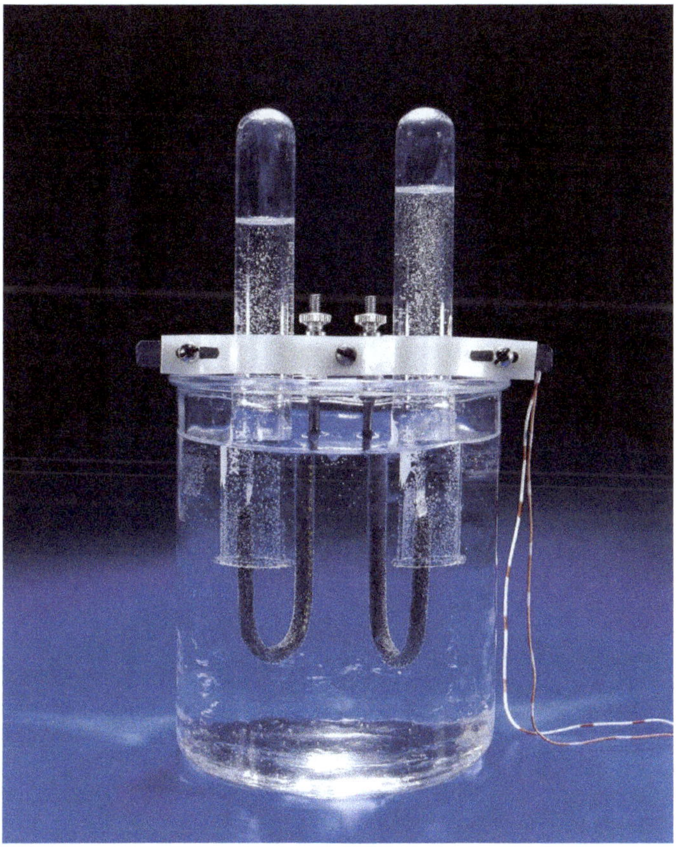

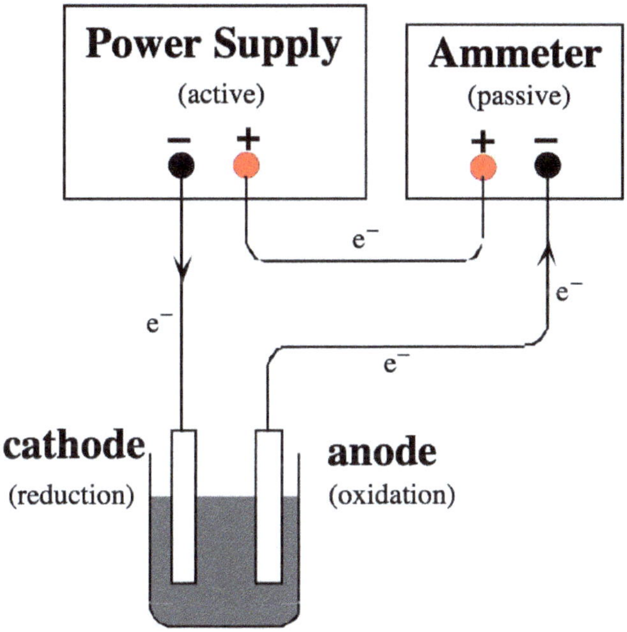

The power supply provides direct current to the cell.

For controlled electrochemical work, an adjustable DC supply offers important advantages because voltage and current can be observed and controlled rather than treated as fixed values.

A supply should also be appropriate for the electrical demands of the cell. Connections should remain secure and protected from accidental contact with conductive liquids or metal objects.

One particularly important distinction must be maintained throughout this book: electrode names are determined by the reactions occurring at them, not simply by wire color.

The anode is where oxidation occurs.

The cathode is where reduction occurs.

This distinction is especially important because informal metal-recovery demonstrations sometimes use the terms inconsistently. A chemist or refiner should identify electrodes according to electrochemical function rather than copying a label without examining the system.

The physical cell should therefore be understood as a combination of chemistry and electrical engineering. The container, electrodes, electrical connections, power source, and electrolyte collectively establish the conditions under which the desired reaction must compete with every other reaction that is possible.

Selecting and Preparing Metal-Bearing Materials

The starting material is one of the most important variables in precious metal recovery.

Gold and silver are rarely encountered as isolated, perfectly clean materials when recovery is the objective. They may exist as thin plating, electrical contact surfaces, coatings, alloy components, powders, residues, or dissolved species within a previously processed solution.

Before choosing an electrochemical process, the refiner must therefore ask a basic question:

Where is the precious metal, and what is it associated with?

Silver-plated copper provides a useful example.

The visible silver-colored surface may represent only a thin layer over a much larger mass of copper. The objective is therefore not necessarily to dissolve the entire object. A more selective process may attempt to separate the valuable surface material while preserving much of the base metal.

Electrical contacts can present a more complicated problem because their construction may include copper or copper alloys, silver-bearing contact material, solder, steel components, and other metals.

Gold-plated electronic components present similar challenges. A visually impressive gold surface does not necessarily represent a large mass of gold. Plating can be extremely thin.

This introduces the concept of **feed material**.

Feed material is the material entering the recovery process. Understanding it helps determine whether electrolysis is appropriate and what kind of electrochemical system would be required.

Material preparation can begin with physical sorting.

Different categories of scrap should generally not be combined merely because they appear to contain the same precious metal. Keeping similar materials together makes the resulting chemistry easier to understand and can reduce unpredictable contamination.

Foreign components may also need to be removed before electrochemical processing.

Plastics, oils, adhesives, dirt, loose corrosion products, and unrelated metal components can interfere with electrical contact or introduce unwanted substances into the system.

However, preparation should not become indiscriminate grinding or shredding.

Reducing material into smaller pieces increases surface area, but it can also mix metals that were previously separated, produce difficult-to-handle particles, and make later mechanical separation more complicated.

The appropriate preparation depends on what the recovery process is intended to accomplish.

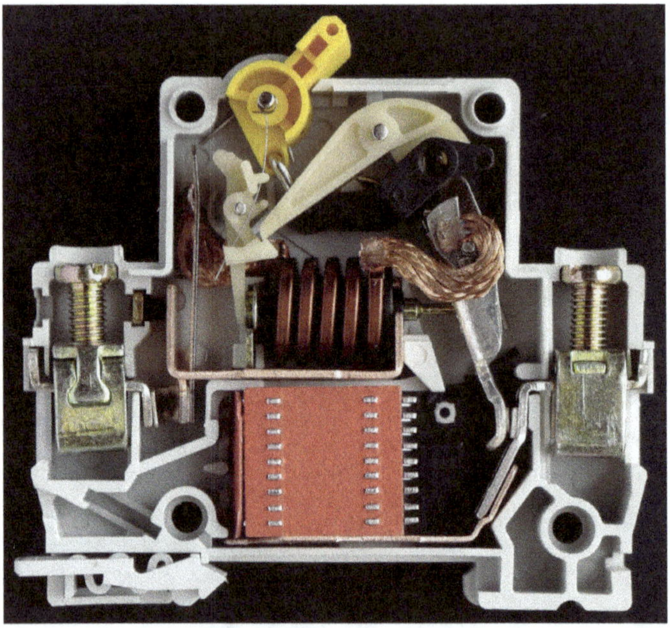

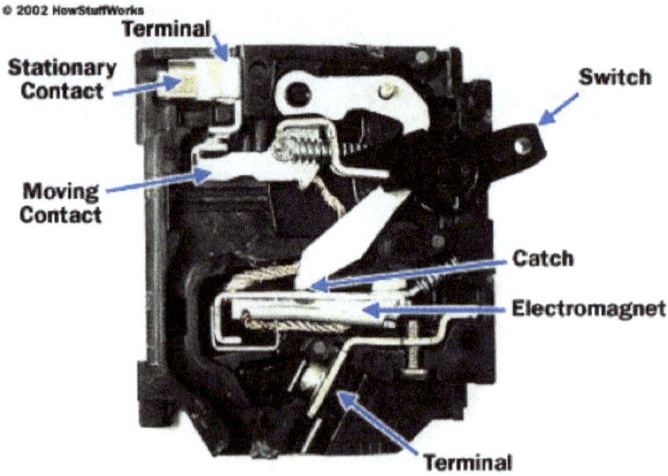

The silver-bearing material used in the reference process demonstrates a comparatively simple preparation strategy. Pieces of silver-plated copper are grouped together and electrically connected so they can participate in the cell as a unit.

That approach also illustrates why exposed conductive surfaces matter.

If electrical contact is unreliable, the process may become uneven. One piece may react rapidly while another receives little effective current. Oxide layers, corrosion, loose connections, and poor physical contact can all influence current distribution.

Surface area must also be considered.

A large amount of material placed into a relatively small cell does not simply behave like a scaled-up version of a smaller batch. Increasing the available reactive area changes the electrical characteristics of the system.

The chemistry of the feed must be considered as well.

Unknown materials deserve particular caution. Scrap components may contain lead, nickel, cadmium, chromium-containing materials, or other metals that introduce chemical, environmental, and health concerns. Electrical and electronic scrap can also contain coatings and nonmetallic materials whose behavior may not be obvious from visual inspection.

This is why identification should come before processing.

A professional approach does not begin with "put it in the cell and see what happens."

It begins by determining what is known about the material, what is unknown, what metal is being targeted, and which contaminants are likely to accompany it.

For chemistry students, this is a useful lesson in experimental design. For refiners, it is fundamental process control.

The better the starting material is understood, the easier it becomes to interpret what happens during electrolysis.

Monitoring Reactions, Deposits, and Changes in the Cell

Once the cell is energized, observation becomes part of the analytical process.

Electrolysis produces visible effects, but those effects must be interpreted carefully. A change in color, formation of bubbles, appearance of sediment, or development of a metallic deposit can provide evidence that reactions are occurring, but appearance alone does not establish exactly which reaction has taken place.

Several observations should be considered together.

Changes at the metal-bearing material may indicate that a coating is being removed, a substrate is being exposed, or an electrode itself is being attacked.

Changes at the opposite electrode may include metallic deposition, discoloration, gas evolution, or accumulation of material on its surface.

The electrolyte may also change.

Color can be particularly informative when transition-metal ions enter solution. Copper-containing solutions, for example, can develop characteristic blue or blue-green coloration depending on their chemical environment. A changing solution color may therefore indicate that the process is affecting more than the precious metal being targeted.

Sediment provides another important observation.

In the silver recovery example, silver-bearing plating material separates from the copper and accumulates at the bottom of the vessel. The exposed copper becomes increasingly visible as processing continues.

That physical change provides a useful indication that separation is occurring.

It does not, however, mean that the sediment should immediately be treated as pure silver.

The collected material can contain copper and other substances, which is why the reference process continues into additional separation and purification stages rather than ending with the initial electrolytic treatment.

This distinction between **recovered material** and refined metal is fundamental.

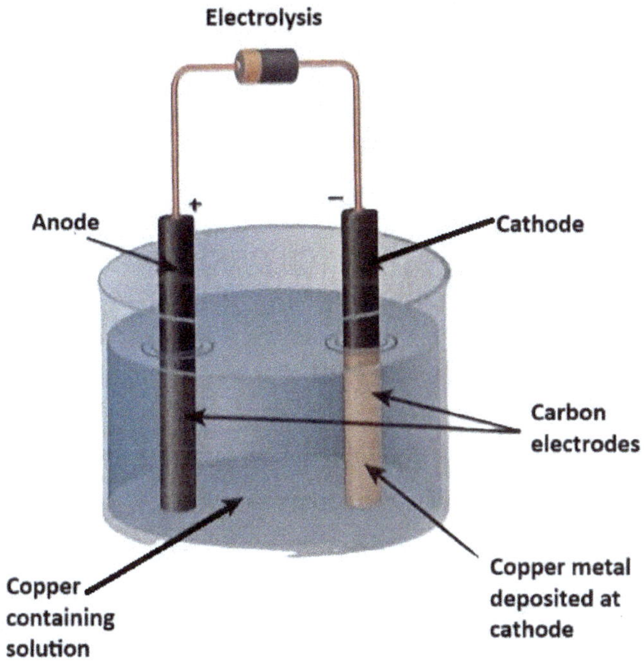
Electrolysis
+
−
Anode
Cathode
Carbon electrodes
Copper containing solution
Copper metal deposited at cathode

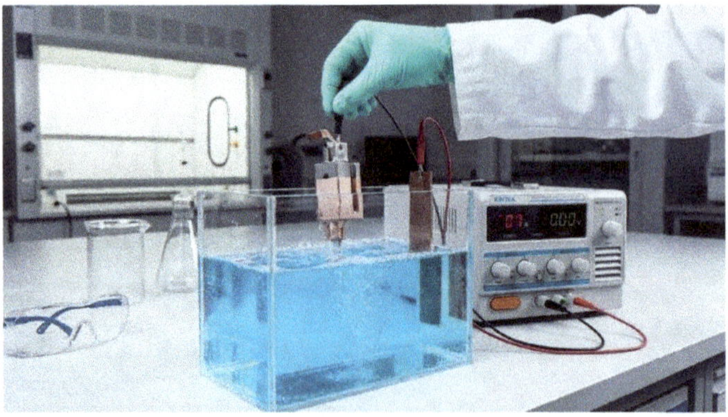

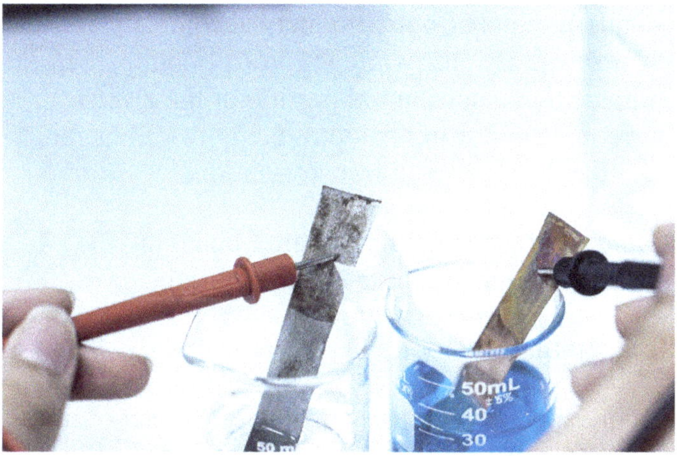

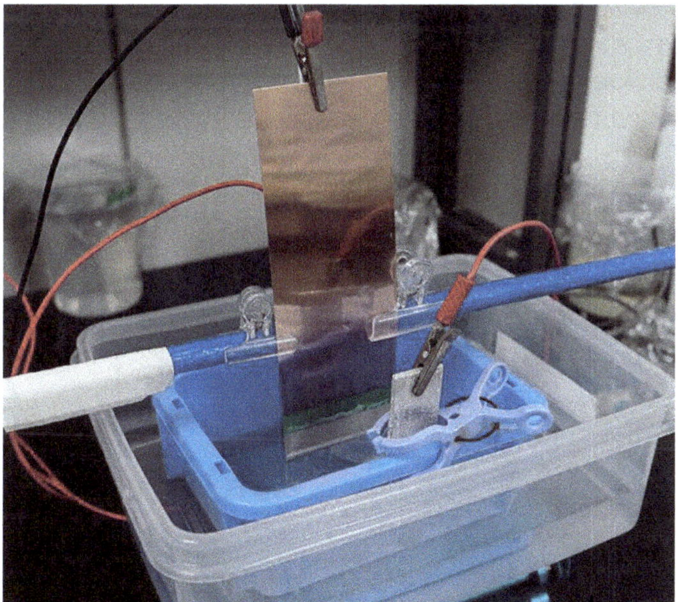

Gas evolution must also be monitored.

Small amounts of bubbling may accompany certain electrode reactions, while vigorous bubbling can indicate that a substantial portion of the electrical current is supporting gas-producing reactions rather than the desired metal transformation.

Rapid increases in bubbling after changing the applied voltage are therefore meaningful.

Temperature is another useful indicator.

An unexpected increase can suggest higher current, greater electrical resistance, accelerated chemical reactions, or other changes in the system. Heating may alter reaction rates and electrolyte behavior, making the cell progressively different from the conditions under which it began.

Voltage and current should consequently be observed alongside physical changes.

If current rises sharply, the operator should not merely celebrate a faster process. The change should be interpreted.

What caused the increase?

Has electrode area changed?

Has conductivity changed?

Has the distance between conductive surfaces changed?

Is another reaction becoming dominant?

These questions turn simple observation into process analysis.

The same reasoning applies to deposits.

A smooth, adherent metallic layer can indicate very different conditions from a loose, dark, powdery deposit. Deposit structure can be affected by current density, concentration, mass transport, electrode condition, temperature, and competing reactions.

In advanced electrochemical work, the appearance of a deposit is therefore information.

It is not final identification.

Confirmation of recovered material may require additional chemical analysis, weighing, purity testing, or further refining.

Monitoring also helps determine when a process should be stopped. Continuing electrolysis after the useful reaction has largely finished may waste energy or encourage unwanted reactions.

The objective is not to run the cell for an arbitrary amount of time.

The objective is to understand what the cell is doing and determine when the intended stage of the recovery process has been completed.

Good records make this easier.

Recording starting mass, electrode dimensions, voltage, current, processing time, temperature changes, observations, recovered mass, and other relevant conditions allows results to be compared between batches.

For chemistry learners, this transforms electrolysis into an experimental system that can be studied quantitatively.

For precious metal refiners, it transforms trial-and-error processing into a repeatable procedure.

That difference becomes increasingly important as we move into the actual recovery of silver.

Last Thoughts

An electrolytic recovery system is simple in appearance but highly dependent on its configuration. Electrode materials, electrical connections, the power supply, electrolyte, starting material, surface area, and cell geometry all influence the reactions that can occur. Just as importantly, the operator must learn to distinguish observation from identification. Bubbles, sediment, exposed base metal, changing solution color, and electrode deposits provide valuable information, but each must be interpreted within the chemistry of the system. Careful material preparation and systematic monitoring create the foundation for repeatable recovery. With the cell itself understood, the next chapter turns to the first complete precious metal application: **electrolytic silver recovery from silver-bearing and silver-plated materials**.

Chapter 3: Electrolytic Silver Recovery

Silver-plated copper and silver-bearing electrical contacts provide an excellent example of how electrolysis can be used as a separation stage in precious metal recovery. The objective is not simply to produce a visible reaction. It is to use electrochemical conditions to separate silver-bearing material from a base-metal substrate while observing what happens to the underlying copper, the electrodes, and the material collected from the cell. This chapter follows the silver recovery workflow from preparation through electrolysis and collection. Because chloride-containing electrolytes and some downstream refining chemistries can generate hazardous gases or involve highly corrosive substances, the process should be understood as controlled laboratory or refining work with appropriate ventilation, chemical controls, and waste handling rather than as an improvised household experiment.

Preparing Silver-Bearing and Silver-Plated Materials

The first step is determining what kind of silver-bearing material is actually being processed.

Silver plating is encountered on electrical contacts, switches, connectors, circuit-breaker components, industrial hardware, and other conductive materials. In many cases, the silver represents only a relatively thin surface layer over a much larger quantity of copper or copper alloy.

That construction determines the recovery objective.

If the silver exists primarily as plating over copper, dissolving or destroying all of the copper simply to obtain the surface silver may be inefficient. An electrolytic separation process can instead be designed around removing the silver-bearing layer while leaving much of the copper substrate available for separate recovery or recycling.

Silver-bearing contacts require additional evaluation because not every silver-colored contact is simply pure silver deposited over pure copper.

Contact materials can contain silver alloys, brazing materials, copper, nickel, tin, and other metals. Some specialized electrical contacts may contain additional components specifically chosen for their electrical and wear characteristics.

The appearance of the material alone is therefore insufficient to determine composition.

Sorting becomes the first practical operation.

Materials of similar construction should be grouped together whenever possible. This makes the behavior of the batch easier to interpret and reduces the number of unknown metals entering the electrochemical system.

Nonmetallic components and unrelated pieces should be removed where practical. Dirt, grease, loose corrosion, plastics, and insulating material can interfere with electrical contact or introduce unnecessary contamination.

Electrical continuity is particularly important.

The reference recovery process groups several pieces of silver-plated copper together and connects them with copper wire. The purpose is straightforward: each piece must become electrically connected to the intended side of the cell if it is expected to participate in the electrochemical process.

A poorly connected piece may remain largely unchanged while neighboring pieces react strongly.

The material should therefore be arranged so that current can reach the conductive surfaces consistently.

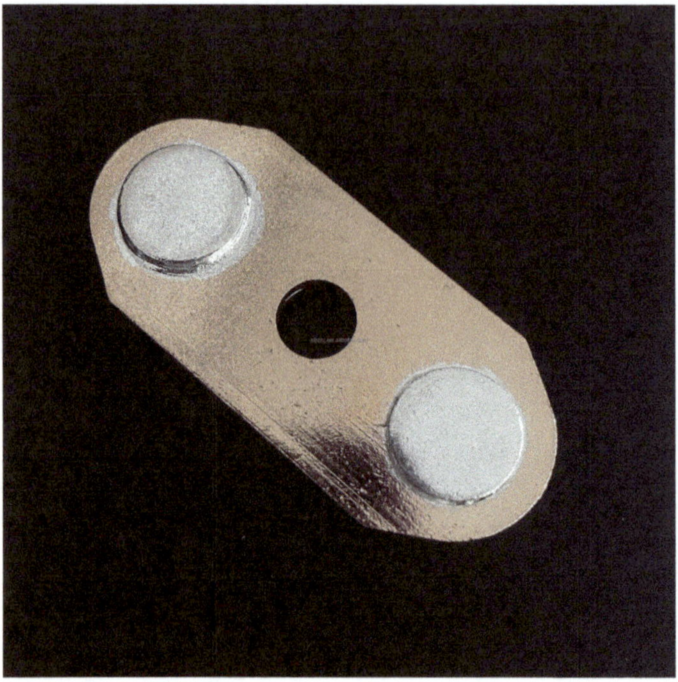

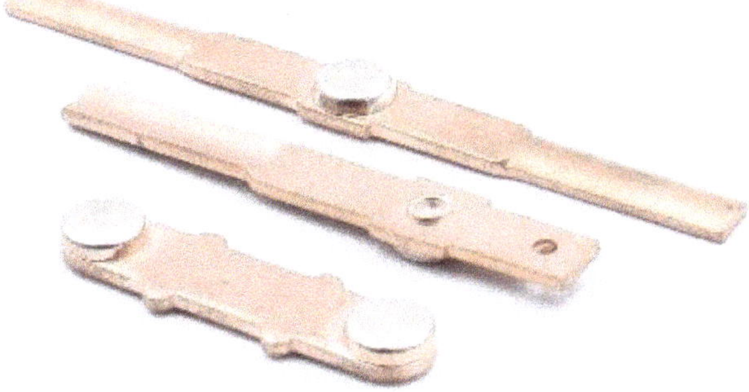

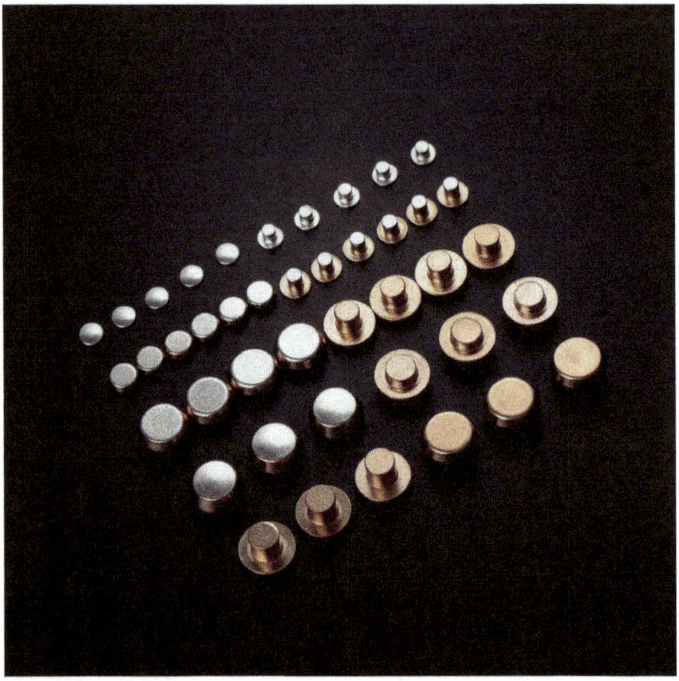

Surface condition also matters.

Heavy contamination or corrosion can affect electrical contact and alter the way current is distributed. Preparation should focus on establishing a known, electrically conductive starting material rather than aggressively reducing everything into small particles.

This is especially important when the base metal has value of its own.

Copper recovered after removal of a silver-bearing surface can remain a recyclable material. Keeping the copper comparatively intact can therefore produce two useful streams: silver-bearing material for further refining and copper-bearing material for separate handling.

Before electrolysis begins, the operator should know at least four things:

What material is being targeted?

Where is the silver located?

What is the principal substrate?

What other metals are likely to be present?

Those questions establish the chemical problem that electrolysis is being asked to solve.

Running the Electrolysis Process Step by Step

The electrolytic stage begins after the material has been sorted, prepared, and electrically connected.

The first operational step is establishing a chemically compatible cell containing the appropriate electrodes and electrolyte for the selected process. The exact electrolyte chemistry is important because the electrolyte is not merely conductive water. Its composition determines which ionic species are available and which competing electrode reactions may occur.

This deserves particular attention in silver stripping systems.

Chloride-containing electrolytes appear in some informal silver-recovery methods, including the source process examined for this book. However, electrolysis of chloride-containing solutions can produce **chlorine gas** under certain electrode conditions. Acidification can further increase the seriousness of the hazard. For that reason, an advanced refiner should not treat a salt-and-acid electrolyte as a casual recipe. Electrolyte selection requires an understanding of electrode reactions, ventilation, compatible materials, and the possible gases generated by the system.

Once an appropriate controlled system has been established, the prepared silver-bearing material is positioned as the electrode at which the intended stripping or oxidation behavior will occur.

The opposing electrode completes the circuit through the electrolyte.

Before energizing the system, electrode spacing and electrical connections are checked. The electrodes must not touch. Conductive wires and clips should remain appropriately positioned, and the power source should initially be off while physical adjustments are being made.

The cell can then be energized under controlled electrical conditions.

This is the point at which observation becomes essential.

The operator watches the silver-bearing surface.

If the process is working as intended, changes should become visible at the interface between the plating and substrate. The silver-bearing layer may loosen, detach, or otherwise separate as the electrochemical process progresses.

The underlying copper becomes increasingly visible.

That physical transformation provides a useful way of following the process.

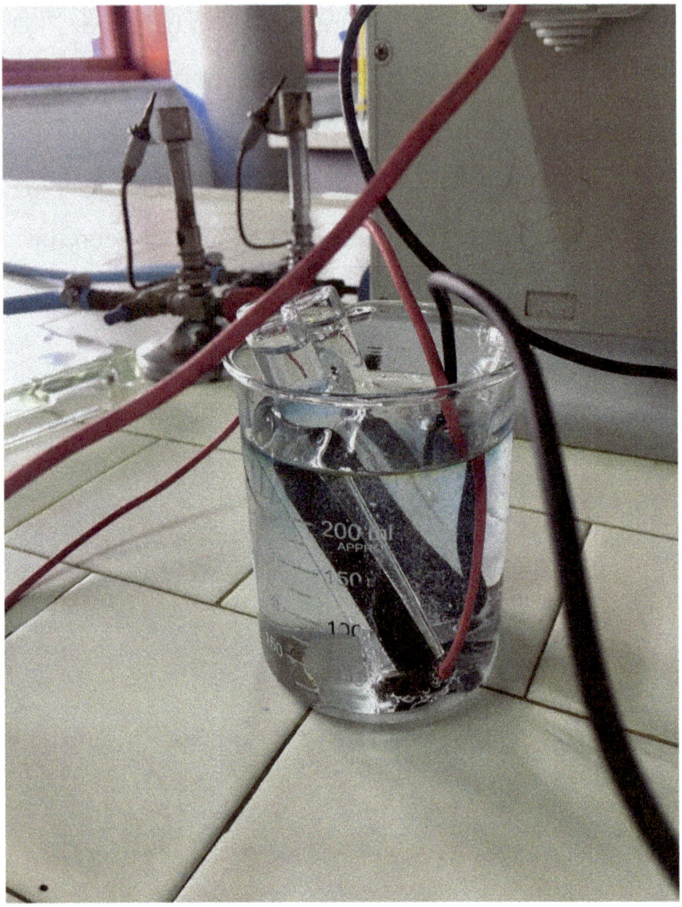
200 ml

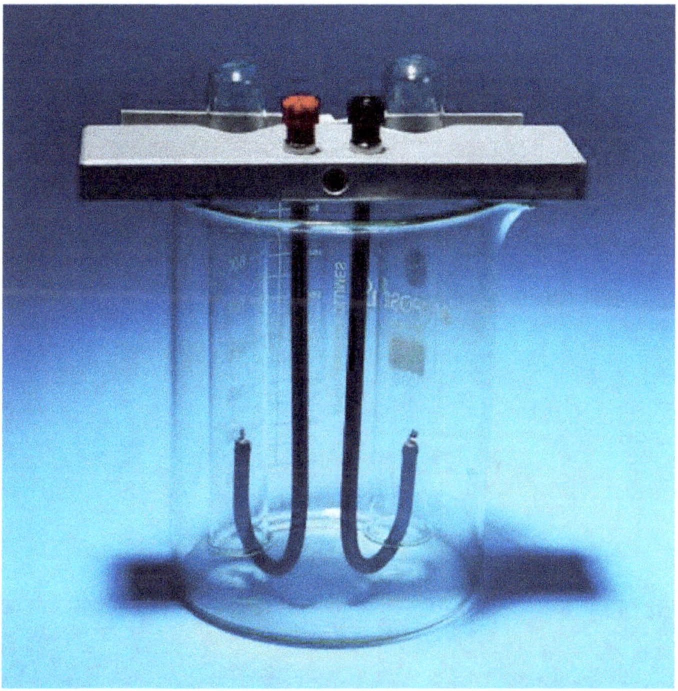

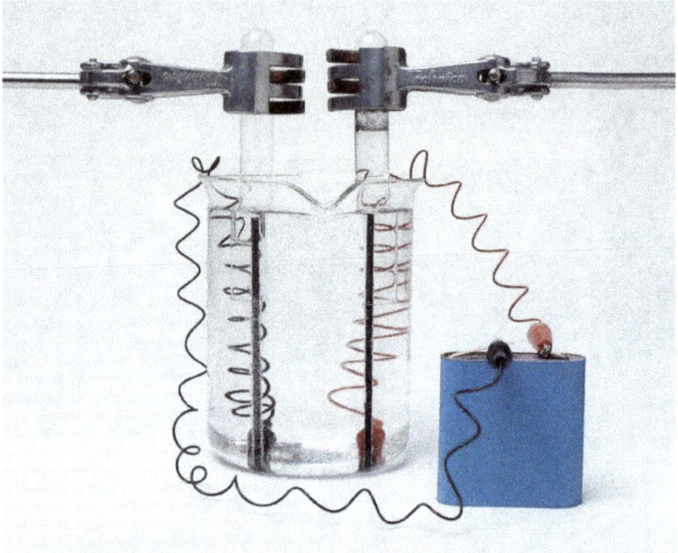

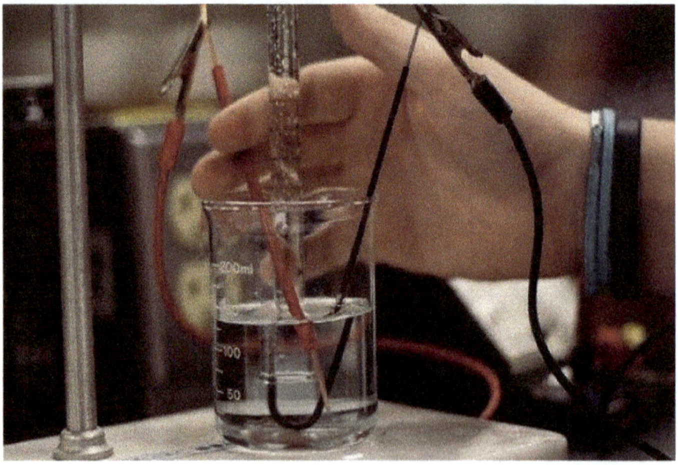
200ml
100
50

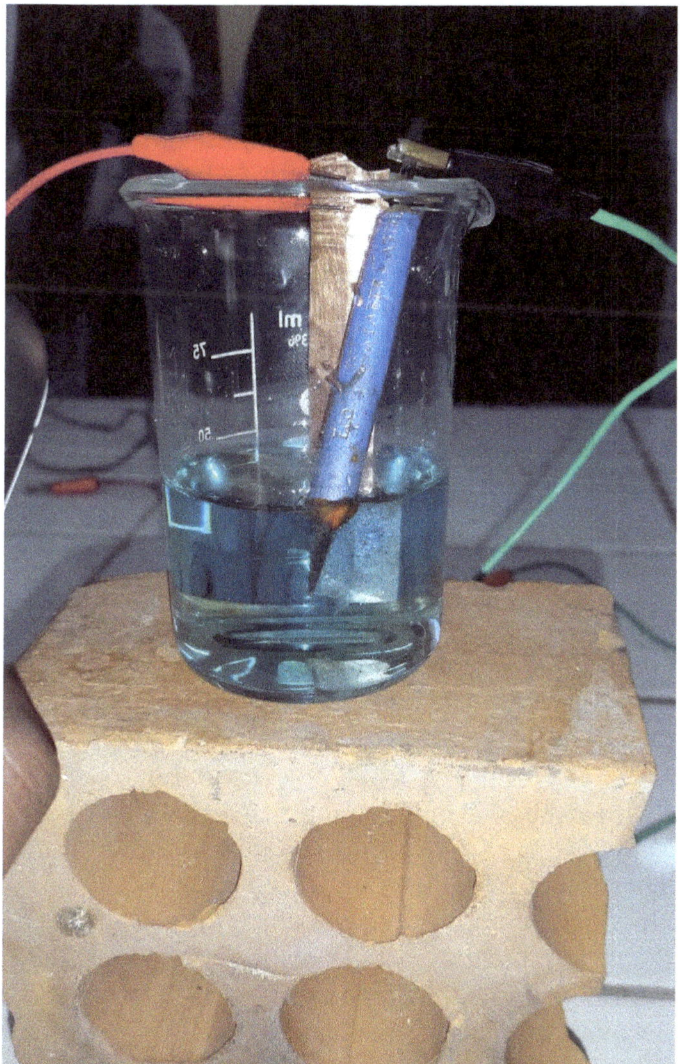

Material may also accumulate elsewhere in the cell.

Some silver-bearing solids can fall away from the substrate and collect as sediment. Other material may adhere temporarily to surfaces within the system.

At the same time, competing reactions must be watched.

Bubbling deserves particular attention. Gas evolution indicates that at least part of the electrical current is driving a gas-producing electrode reaction. Increasing the electrical input simply to accelerate stripping can intensify those reactions and can also increase heating.

This is why the process should not be approached as:

More voltage = faster recovery = better recovery.

Instead, the objective is controlled separation.

The voltage, resulting current, electrode area, electrolyte composition, and changing condition of the material all interact. As plating is removed and different surfaces become exposed, the electrical behavior of the cell can change.

The substrate itself must also be monitored.

If copper begins participating significantly in unwanted reactions, the recovered solids can become increasingly contaminated with copper-bearing material. That creates more work during the purification stage and can reduce the efficiency of the overall process.

A useful electrochemical recovery process therefore stops being judged merely by how quickly the original material changes appearance.

The better question is:

How selectively is the desired separation occurring?

As the silver-bearing surface is removed, exposed copper provides a visual indication of progress. Pieces can be examined to determine whether substantial plated areas remain.

Processing should not continue indefinitely after the useful separation has been achieved. Once the intended stripping stage is complete, unnecessary electrolysis can promote additional reactions without producing meaningful additional silver recovery.

Power is switched off before the cell is physically disturbed.

This sequencing is important.

Electrical power should not remain applied while electrodes, conductive tools, or metal-bearing materials are being repositioned or removed.

The processed material can then be withdrawn and inspected.

At this point, two distinct products may be present.

One is the exposed copper-rich substrate.

The other is the collected silver-bearing material.

Neither should be confused with the original feed.

The separation has changed the material streams, but the precious-metal stream has not necessarily reached final purity.

Collecting and Evaluating the Recovered Silver-Bearing Material

Material collected after electrolysis may look like the desired product, but appearance alone cannot establish purity.

This is one of the most important distinctions in the entire recovery workflow.

The solids recovered from the cell should initially be regarded as **silver-bearing concentrate** rather than automatically labeled refined silver.

They may contain detached silver plating, copper particles, other contact metals, oxides, salts, electrolyte residues, and material released from electrodes or electrical connections.

Collection therefore needs to preserve as much of the valuable fraction as possible while preparing it for evaluation and subsequent purification.

Sediment should be allowed to settle where appropriate before separation from the bulk liquid. Fine particles can remain suspended considerably longer than larger pieces, so rushing this stage can result in valuable material being carried away with the liquid fraction.

Material attached to processed copper also deserves attention.

Loose silver-bearing particles may remain trapped between pieces, cling to surfaces, or collect in corners of the vessel. Careful rinsing and consolidation can reduce these mechanical losses.

The recovered copper-rich material should be kept separate from the silver-bearing solids.

This is useful economically and analytically.

If the stripping stage has worked well, much of the original copper remains identifiable as a separate base-metal stream rather than being unnecessarily mixed into the precious-metal concentrate.

MM
CM 1 2 3 4 5 6 7 8

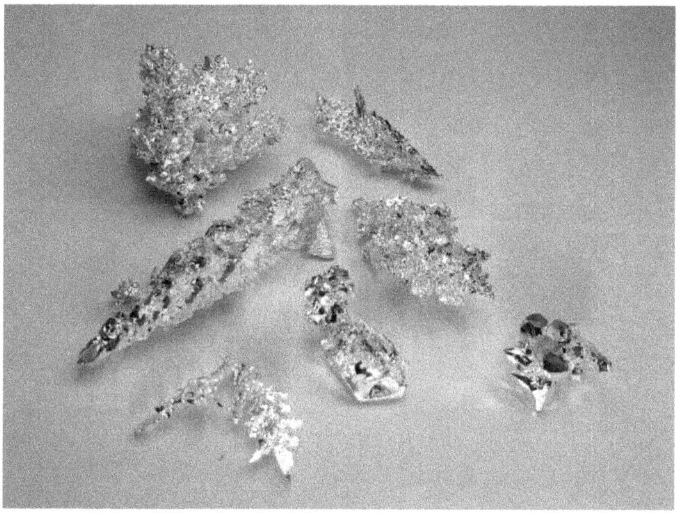

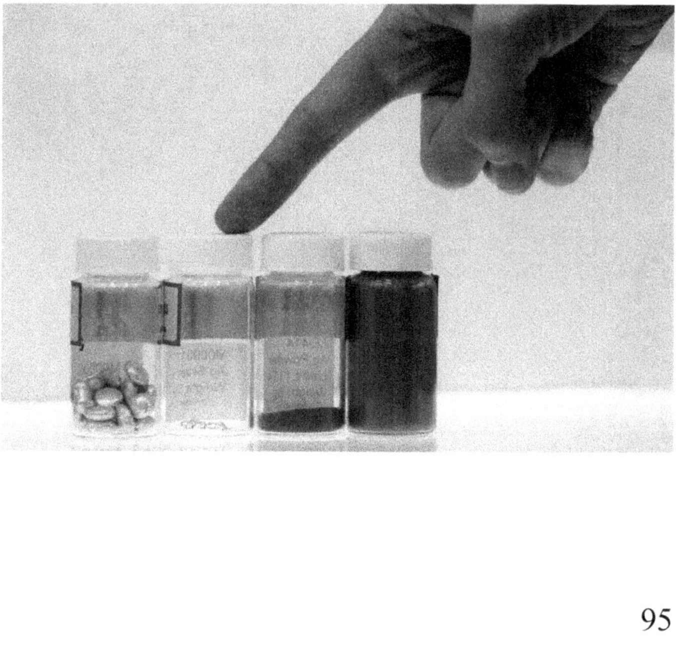

SILVER

The collected silver-bearing material can then be evaluated.

Color and texture provide observations, but they should not be treated as definitive identification.

A gray or metallic-looking powder is not automatically silver.

A dark deposit is not automatically high-purity precious metal.

Likewise, a shiny particle does not prove that a refining stage has been successful.

Mass provides another useful measurement.

The original feed material can be weighed before processing, the recovered copper-rich substrate can be weighed afterward, and the collected concentrate can eventually be weighed after appropriate drying and processing. These measurements help establish a **mass balance**.

A simplified mass balance asks where the starting material went.

If the starting batch loses a measurable amount of mass, some of that material should appear in the recovered solids, solution, electrode deposits, or other process streams.

Mass accounting cannot by itself identify the metals, but it can reveal losses and inconsistencies.

For example, a surprisingly large mass of recovered "silver" relative to the amount of plating expected on the feed material is a warning sign.

The material probably contains something besides silver.

Copper contamination is particularly important because silver-plated copper places the valuable layer directly against a much larger reservoir of base metal. Excessive attack on the substrate can introduce copper into the recovered fraction.

This is where the chemistry student and the refiner arrive at the same conclusion from different directions.

The chemistry student recognizes that multiple oxidation and reduction reactions may have occurred.

The refiner recognizes that the collected material still requires purification.

The electrolysis stage has accomplished **separation**, but separation and purification are not synonymous.

This is also why immediately melting an unknown recovered concentrate can be misleading. Melting consolidates the material, but it does not automatically remove every contaminant or establish its composition. A metallic button produced from impure feed remains an impure metallic button.

The recovered silver-bearing material should instead move into a deliberate purification workflow.

That workflow begins by determining which metals and residues are likely to be present, separating the desired silver from those contaminants, and converting the recovered silver into a form that can be cleaned and ultimately returned to metallic silver.

That is the subject of the next chapter.

Last Thoughts

Electrolytic silver recovery demonstrates why precious metal refining is best understood as a sequence rather than a single operation. The starting silver-bearing material must first be identified and prepared, the electrochemical system must then produce a controlled separation, and the resulting material must finally be collected without assuming that it is already pure. Exposed copper, detached plating, sediment, electrical behavior, and changes within the cell all provide information about the progress of the process. The most important result of this stage is therefore not simply "silver at the bottom of the vessel." It is the creation of a more concentrated **silver-bearing material stream that can now be separated from contaminants and refined further**. Chapter 4 follows that material through the next stage: purifying and refining recovered silver.

Chapter 4: Purifying and Refining Recovered Silver

The material collected after electrolytic stripping should not automatically be considered pure silver. It may contain silver along with copper, residues from the electrolyte, particles from the original contacts, and other contaminants introduced during recovery. The next objective is therefore to move from **recovery to purification**. This chapter follows the recovered silver-bearing material through the major stages demonstrated in the recovery workflow: separating silver from copper and other contaminants, converting the silver into forms that can be cleaned and isolated, and finally evaluating the metallic silver produced. Some of these operations can involve highly corrosive reagents, heat, reactive solutions, and hazardous fumes, so the chemistry is presented as an advanced refining workflow rather than a casual household procedure.

Separating Silver From Copper and Other Contaminants

The first challenge is determining what remains mixed with the recovered silver.

When silver plating is separated from a copper substrate, some copper can accompany the recovered material. Small fragments can detach mechanically, while electrochemical reactions can introduce additional copper-containing material into the collected fraction.

The recovered solids may therefore represent a mixture rather than a single metal.

This becomes visible during subsequent chemical processing. In the reference workflow, the recovered plating is subjected to a dissolution stage, after which the resulting solution develops a blue coloration associated with dissolved copper species.

This is an important analytical observation.

The color demonstrates that copper has traveled beyond the original substrate and entered the refining stream. The initial electrolytic separation therefore did not produce pure silver.

That is not necessarily a failure.

A recovery stage is often designed to concentrate the valuable material enough that subsequent purification becomes practical.

The important point is recognizing what each stage has accomplished.

After dissolution, insoluble material can be separated from the liquid through **filtration**.

Filtration provides a physical separation between solids that remain undissolved and chemical species present in the liquid phase.

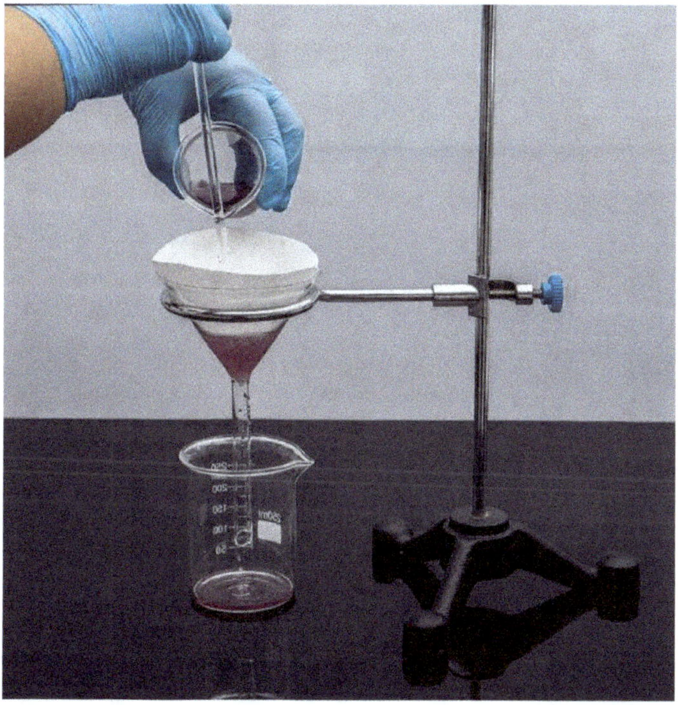

100ml
IWAKI
100
80
60
40
20

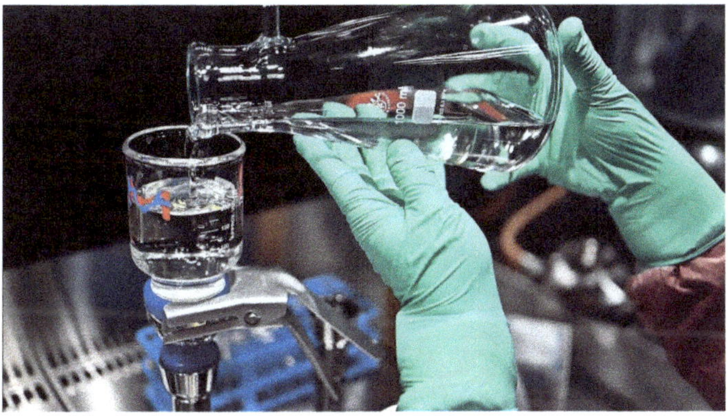

At this point, the silver is no longer being handled simply as pieces of metallic plating. It is participating in solution chemistry.

This transition is important because dissolved metals can be separated according to differences in their chemical behavior.

Silver has a particularly useful relationship with chloride ions. Under appropriate conditions, dissolved silver ions react with chloride ions to form **silver chloride**, an insoluble compound.

The simplified reaction can be written:

$Ag^+ + Cl^- \rightarrow AgCl(s)$

The "(s)" indicates that silver chloride forms as a solid.

This is a precipitation reaction.

Instead of remaining dissolved throughout the liquid, the silver becomes part of a solid phase that can settle and eventually be separated.

Copper behaves differently under these conditions. Much of the dissolved copper can remain in the liquid while silver chloride forms a separate solid phase.

That difference provides a route to chemical separation.

However, precipitation should not be confused with producing metallic silver.

Silver chloride contains silver, but it is a silver compound.

The silver still needs to be converted back into its elemental metallic form.

This is why the refining process continues.

Understanding Precipitation, Conversion, and Purification

Precipitation is one of the most useful concepts connecting analytical chemistry with precious metal refining.

A dissolved metal exists as ions or related dissolved species distributed throughout a solution. If those species react to produce a sufficiently insoluble compound, the metal can move from the liquid phase into a solid phase.

Silver chloride provides a classic example.

When chloride is introduced into a suitable solution containing silver ions, a white silver chloride precipitate can form.

The appearance of the precipitate provides useful evidence that silver-containing material is being transferred into a solid form, but careful processing is still required.

Fine precipitated particles may remain suspended for some time.

Allowing the material to settle helps separate the silver-containing solid from the liquid above it. Removing the liquid too aggressively can carry fine silver chloride particles away and reduce recovery.

The precipitate can then be washed.

The reference workflow repeatedly uses distilled water during washing. The purpose is not to dissolve the silver chloride. Instead, washing helps remove soluble contaminants that remain trapped within or around the precipitated material.

This illustrates an important refining principle:

Purity can improve through repeated separation of the desired material from unwanted soluble substances.

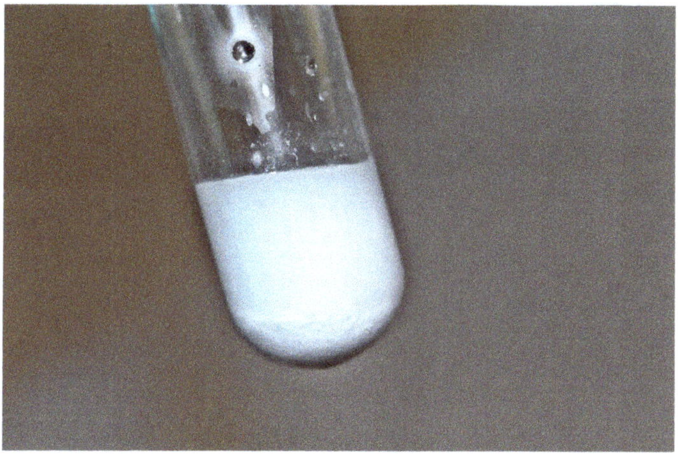

50
30

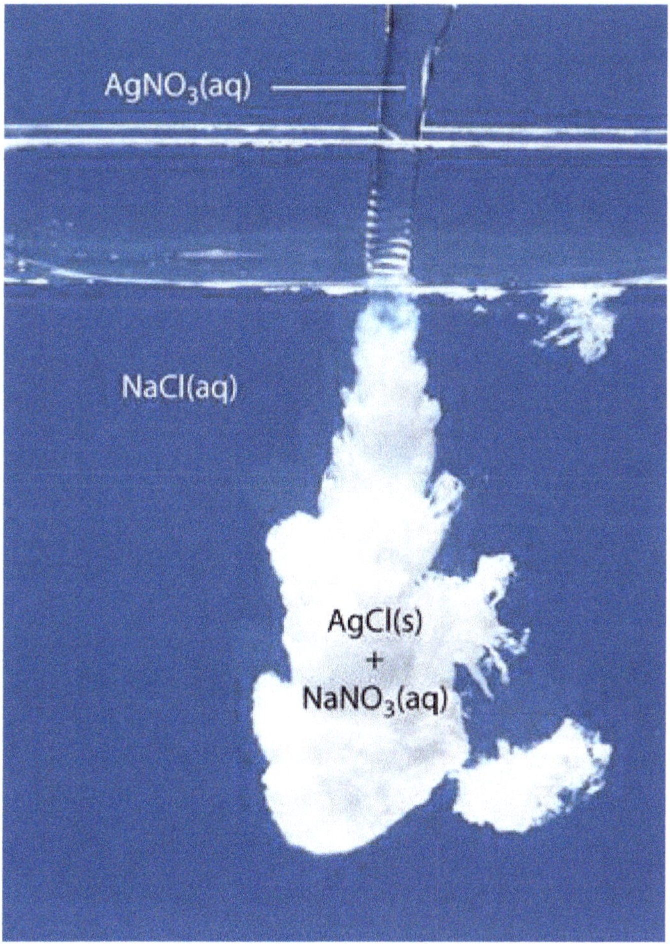
$AgNO_3(aq)$
$NaCl(aq)$
$AgCl(s)$
+
$NaNO_3(aq)$

Silver chloride itself has distinctive behavior.

Exposure to light can cause it to darken as photochemical reactions occur. For chemists, this is a familiar property of silver halides and historically contributed to their importance in photographic chemistry.

For the refiner, the more immediate concern is converting the purified silver-containing compound back toward metallic silver.

The reference process does this through an intermediate conversion.

The washed silver chloride is treated so that it is converted into a dark silver oxide-containing material. This produces a visible transformation from the lighter chloride precipitate toward a darker solid.

A simplified representation of the conversion can be written as:

$$2AgCl + 2OH^- \rightarrow Ag_2O + 2Cl^- + H_2O$$

The resulting **silver oxide** is still not metallic silver.

Another reduction step is required.

In the demonstrated workflow, a reducing sugar is used to convert the silver-containing oxide into metallic silver. Chemically, the important concept is reduction: silver in the compound gains electrons and is returned toward the elemental metallic state.

The visual appearance changes again.

Instead of treating these color changes as isolated tricks, it is useful to follow the silver through the entire sequence:

Metallic silver-bearing material -> dissolved silver species -> silver chloride -> silver oxide -> metallic silver

This sequence is one of the clearest examples in the book of how a metal can move between different chemical forms while remaining the element of interest.

The silver has not disappeared when it enters solution.

It has changed chemical form.

Likewise, precipitating silver chloride does not create new silver. It transfers existing silver into a different compound that is easier to separate from portions of the solution.

The final reduction does not manufacture silver either. It converts silver already present in the compound back into elemental metal.

This perspective is useful for chemistry students because it connects oxidation states, solubility, precipitation, and reduction into one continuous process.

It is equally useful for refiners because each chemical transformation has a practical separation purpose.

After reduction, washing becomes important again.

The resulting metallic silver can retain soluble reaction products and other residues. Repeated washing helps remove those materials before the silver is dried and evaluated.

This is another reason refining should be understood as a sequence of controlled transformations rather than a single dramatic reaction.

Each stage prepares the material for the next.

Producing and Evaluating Refined Silver

After the silver-containing compound has been reduced, the recovered material can appear as a fine metallic powder or granular mass rather than the familiar bright solid associated with finished silver products.

That appearance is normal for finely divided metal.

The material should first be thoroughly separated from the remaining liquid and washed as required by the refining method. Water and other volatile material must be removed before any high-temperature operation is considered.

The dry metallic material can then be consolidated.

Melting changes the physical form of the recovered silver by bringing separate particles together into a larger metallic mass. It can make weighing, handling, and later evaluation easier.

However, melting should not be mistaken for a complete purification method.

If copper or another metal remains mixed with the silver, simply melting the material can produce an alloy containing both.

Likewise, residues carried into the melt can produce slag, oxides, or inclusions.

The purity of the feed entering the melting stage therefore matters.

5mm
1/4"

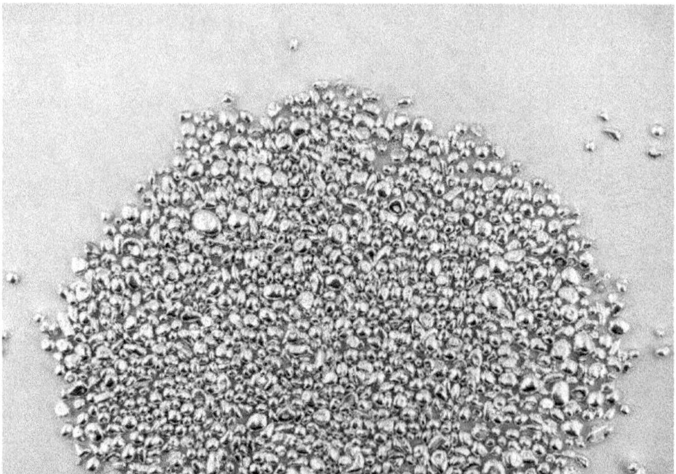

The reference process ultimately melts the recovered material and obtains a consolidated silver product. The example demonstrates the complete progression from plated scrap through electrolytic separation, chemical purification, reduction, washing, and final consolidation.

For an advanced refiner, however, producing a metallic button is not sufficient evidence of purity.

Evaluation should compare what was expected with what was actually recovered.

Mass is one of the simplest measurements.

If the starting material was weighed and the intermediate streams were tracked, the final silver mass can be compared with estimates of how much silver the feed material was likely to contain.

A recovery calculation can be expressed generally as:

Recovery (%) = (Recovered Silver / Estimated Silver in Feed) x 100

This number must be interpreted carefully.

If the recovered material contains contaminants, its mass can make apparent recovery look artificially high.

For example, suppose the estimated feed contains 20 grams of silver and the final metallic product weighs 22 grams. That does not demonstrate 110 percent silver recovery.

It suggests that the estimate, measurement, or purity assumption is incorrect.

The product may contain copper or other material.

This illustrates why **recovery percentage and purity are different measurements**.

A process can recover nearly all of the available silver while producing a relatively impure product.

Another process might produce extremely pure silver while losing a significant portion of the available silver during processing.

A successful refining workflow attempts to balance both.

Professional evaluation may use analytical techniques capable of determining composition more directly. Depending on the required accuracy and available equipment, refiners may use methods such as X-ray fluorescence, fire assay, instrumental analysis, or other appropriate testing.

Density and visual appearance can provide supporting information but should not substitute for reliable compositional analysis when purity matters.

This is particularly important when silver is being represented commercially at a specific fineness.

A shiny metallic product is not automatically .999 silver.

Purity is a measurement, not an appearance.

The final material should therefore be understood in terms of the complete process history: what entered the system, what was removed, what remained in solution, what was precipitated, what was washed away, what was reduced, and what ultimately entered the melt.

That material accounting turns silver refining from a sequence of visual reactions into a controlled chemical process.

Last Thoughts

Purifying recovered silver demonstrates how electrochemistry and solution chemistry can operate as complementary parts of the same refining workflow. Electrolysis performs the initial separation, but the collected material may still contain copper and other contaminants. Dissolution, filtration, precipitation, washing, chemical conversion, reduction, and final consolidation progressively separate the silver from those unwanted materials. Following silver through these different chemical forms also reveals an important principle: **recovery is about accounting for the element as it moves from one form and process stream to another**. A final metallic product should never be assumed pure simply because it looks like silver. Its history, mass balance, and composition determine what has actually been recovered. With the silver workflow complete, the next chapter applies the same electrochemical reasoning to a different precious metal and a different starting system: **electrolytic gold recovery**.

Chapter 5: Electrolytic Gold Recovery

Gold recovery by electrolysis differs from the silver-plating process examined in the previous chapters. Instead of beginning with silver attached directly to a copper substrate, electrolytic gold recovery can begin with gold already present in a prepared solution after earlier processing has transferred it from the original material into a dissolved chemical form. Electrolysis is then used as a recovery stage that encourages dissolved gold species to undergo reduction and form a collectible solid. This makes gold recovery an especially useful example of how electrode reactions, voltage, time, solution composition, and competing reactions work together in an electrochemical system.

Understanding Gold-Bearing Solutions and Electrochemical Behavior

Before gold can be recovered electrolytically from a solution, it must already exist there in an appropriate dissolved chemical form.

This distinction is important.

Placing a piece of gold-plated electronic scrap into an arbitrary electrolyte is not equivalent to performing electrolytic recovery from a prepared gold-bearing solution. Earlier processing may be necessary to separate the gold from its original substrate and transfer it into solution.

Once dissolved, gold no longer exists as the familiar yellow metallic surface visible on jewelry, contacts, or plated components. It exists as dissolved gold-containing species whose exact chemical form depends on the solution chemistry used to prepare them.

For electrochemical recovery, the objective is to move gold from that dissolved state back toward elemental metal.

At the cathode, an appropriate gold-containing species can accept electrons and undergo reduction. A simplified conceptual representation is:

$$Au^{n+} + ne^{-} \rightarrow Au$$

This equation intentionally represents the process in general terms. The actual gold species and corresponding half-reaction depend on the chemical system.

That distinction matters because the electrolyte is not simply a passive liquid carrying gold from one place to another.

Its composition affects the stability of dissolved gold species, conductivity, electrode behavior, competing reactions, and the conditions under which gold reduction can occur.

100
ГОСТ 25336-82
100 мл
ТС
ОРИЕНТИРОВОЧНЫЙ ОБЪЕМ
50

79
Au
GOLD
79
Au
GOLD
196.97

METABISULFITE
DE
SOUDE
1KG
150
100
50
75

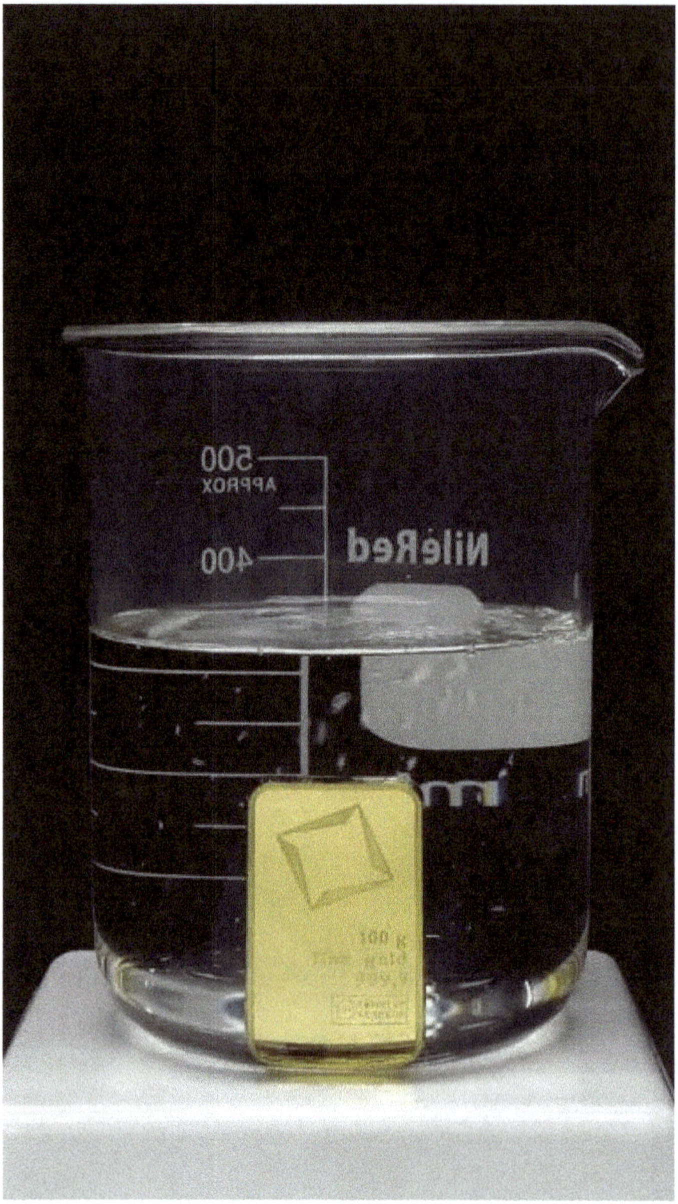
500
APPROX
400
NileRed

Gold's reputation as a relatively unreactive noble metal can sometimes create the impression that its electrochemistry should be simple. In reality, the behavior of gold in solution depends strongly on its chemical environment.

A gold-bearing solution can also contain other dissolved metals.

If the original material included copper, nickel, silver, or other base metals, earlier processing may have introduced some of those elements into the solution as well. The electrochemical system must therefore be understood as a mixture of potentially reducible species rather than a container holding only gold.

This brings selectivity back into the recovery process.

Ideally, conditions favor reduction of the desired gold species while limiting unwanted deposition and other electrode reactions.

The reference gold-recovery demonstration begins with an already prepared gold-bearing solution. It also notes that residual nitric acid is dealt with before electrolysis. The source does not provide enough detail to reconstruct the preparation of that solution or establish its exact composition, so this chapter does not treat that missing preparation as a defined recipe.

That limitation itself illustrates an important principle.

A gold electrolysis procedure cannot be separated from knowledge of the solution entering the cell.

If the composition is unknown, predictions about electrode reactions become much less reliable.

For a chemist, the solution is a collection of chemical species with particular reduction possibilities.

For a refiner, it is a process stream containing gold that must be recovered without unnecessarily carrying other materials into the final product.

Both perspectives require the same thing: understanding what is actually in the solution before attempting to control what comes out of it.

Running the Gold Electrolysis Process Step by Step

The gold recovery demonstration used as a reference employs two stainless steel electrodes placed in the prepared gold-bearing solution and connected to an electrical supply.

The source emphasizes relatively low voltage.

It reports that a reference value of approximately 0.5 to 1 volt could require a process lasting as long as several days, while the demonstrated process increases the applied voltage and reports completing the electrolysis in approximately five hours.

These observations are useful because they illustrate a tradeoff that appears repeatedly in electrochemistry.

Increasing the electrical driving force can accelerate observable reactions.

It can also change which reactions occur.

The objective is therefore not simply to select the highest voltage that produces visible activity.

Before power is applied, the electrodes should be positioned without touching one another, the electrical connections should be established correctly, and the solution and equipment should be treated according to the hazards associated with their actual chemical composition.

Once energized, the cathodic region becomes particularly important because this is where reduction occurs.

Gold-containing species reaching an appropriate cathodic surface can gain electrons and move toward elemental gold.

However, gold deposition may compete with other reduction reactions.

Hydrogen evolution can occur in aqueous systems under suitable conditions. Other dissolved metal ions may also be reduced if the electrode potential permits it.

This is why **electrode potential** is central to selective electrochemical recovery.

Different reactions become favorable under different electrochemical conditions. Controlling those conditions helps determine whether the process preferentially recovers gold or produces a mixture of reactions.

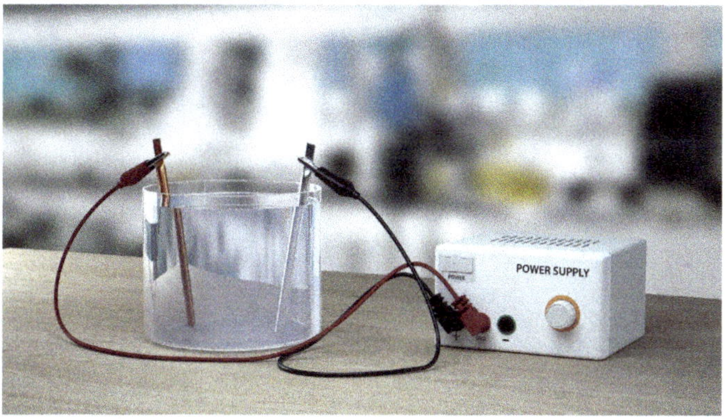
POWER SUPPLY

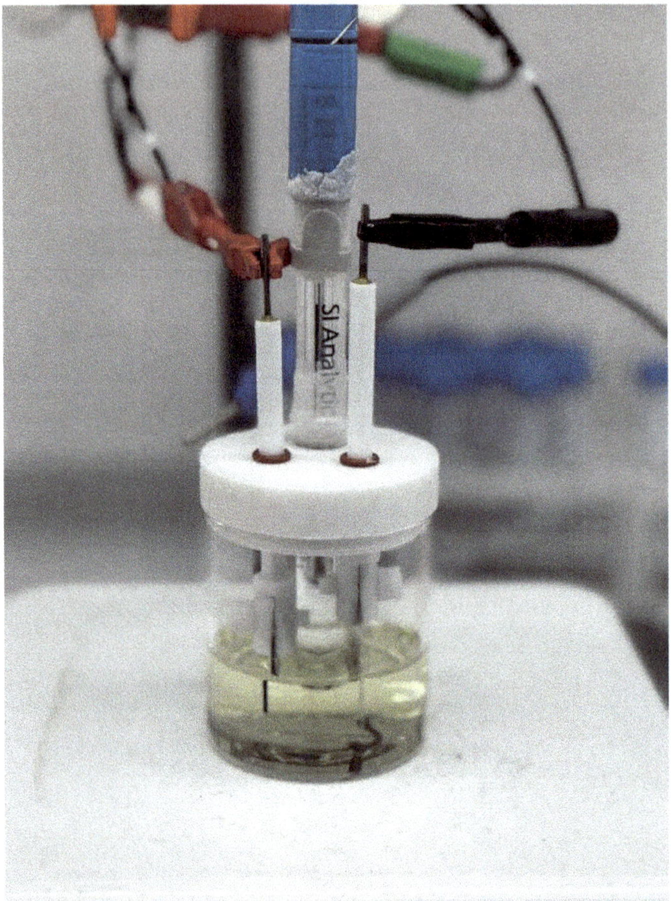

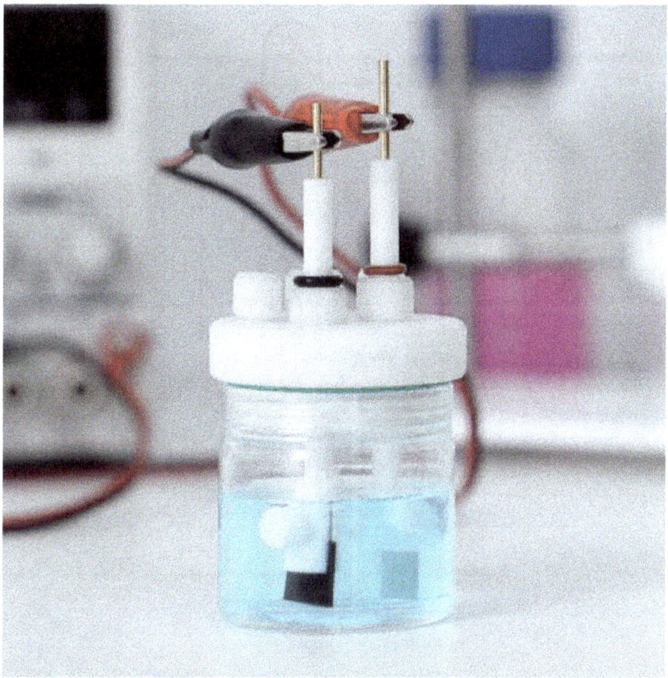

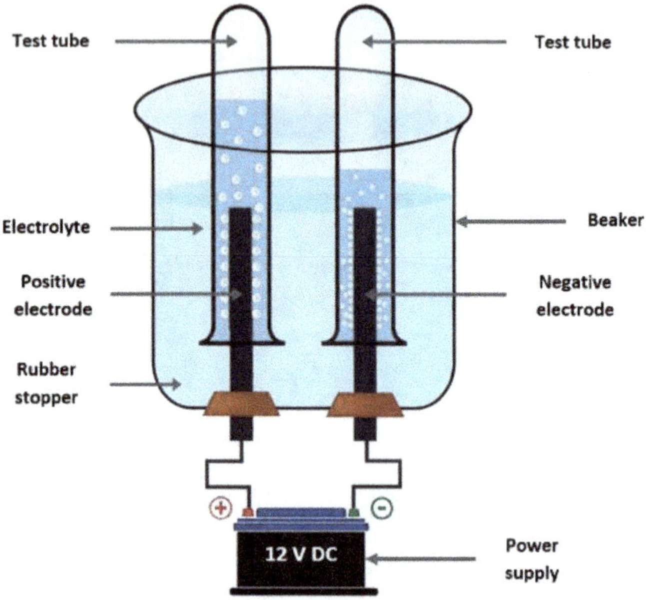

The process should be monitored as it progresses.

Changes in current can indicate that conditions within the cell are changing. Gas evolution can indicate competing reactions. Deposits may appear on electrode surfaces, while particulate material can also accumulate elsewhere in the system depending on the character of the deposit.

The reference demonstration describes the recovered gold as a dark sediment with an appearance compared informally to chocolate spread.

This is a useful reminder that freshly recovered gold does not necessarily resemble a polished gold bar.

Finely divided metallic particles interact with light differently from a smooth bulk surface. A dark brown, black, or otherwise unexpected appearance therefore does not automatically rule out the presence of gold.

At the same time, appearance does not prove that the material is gold.

The chemistry and subsequent evaluation must support that conclusion.

Time is another important process variable.

As dissolved gold is removed from solution, its concentration changes. The conditions that existed at the beginning of the run therefore do not necessarily remain constant throughout the entire process.

Mass transport becomes increasingly important as concentration decreases.

Gold-containing species must reach the electrode surface before they can participate in the electrode reaction. Their movement can involve diffusion, migration, and fluid motion within the cell.

This introduces **mass transport** as another limitation on recovery rate.

Increasing voltage cannot indefinitely compensate for limited transport of the desired species to the electrode. If gold species are not arriving at the reaction surface fast enough, additional electrical driving force may increasingly support other reactions instead.

The process should therefore be viewed dynamically.

Early in the run, gold concentration may support relatively rapid recovery.

Later, as the available gold decreases, the same operating conditions may become less selective or less efficient.

This is one reason a controlled process pays attention to the electrical response and chemical state of the system rather than relying only on a predetermined running time.

When the intended electrolysis stage is complete, power is switched off before the electrodes or other conductive components are handled.

The resulting material can then move into collection.

Collecting, Washing, and Evaluating Recovered Gold

Electrolytically recovered gold may appear as an adherent electrode deposit, loose particulate material, sediment, or a combination of forms depending on the electrochemical conditions.

Every potentially gold-bearing fraction therefore needs to be considered during collection.

Material attached to an electrode should not be ignored simply because sediment is visible elsewhere in the cell. Likewise, fine particles suspended in the liquid can represent material that has not yet settled.

Allowing fine solids to settle can help reduce mechanical losses.

The bulk liquid can then be separated carefully from the recovered solids using an appropriate method for the process.

Washing removes soluble electrolyte residues and other substances associated with the liquid phase.

As with silver recovery, washing is not merely cosmetic. Residual soluble material can contaminate the recovered product, affect its measured mass, and interfere with later evaluation.

After washing and appropriate drying, the recovered material can be examined more meaningfully.

Mass should be recorded.

If the amount of gold originally present in the solution is known through analysis or reliable process accounting, the recovered mass can contribute to an estimate of recovery efficiency.

The same caution discussed for silver applies here.

Recovered mass is not automatically pure gold mass.

Other metals can codeposit. Insoluble contaminants can become trapped in the recovered material. Residual salts can increase apparent mass if washing and drying are incomplete.

A material that contains gold is therefore not necessarily refined gold.

This distinction becomes increasingly important as the target purity rises.

For example, material containing a high percentage of gold may be an excellent recovery product while still requiring additional refining before it can be represented as high-purity bullion.

The appearance of the material also changes when fine gold particles are consolidated.

Finely divided recovered gold can look dark and unimpressive compared with familiar polished metal. Once sufficiently purified and consolidated, its characteristic metallic appearance becomes much more recognizable.

But consolidation should come after adequate purification and evaluation.

Melting an impure recovered powder can produce a gold-bearing alloy without resolving the underlying contamination.

Analytical testing provides stronger evidence.

Depending on the required accuracy, refiners can use suitable compositional methods to determine how much gold and what other metals are present. This is especially important when comparing different electrolysis conditions.

A refiner might obtain a greater mass under one set of electrical conditions but discover that the material contains more contamination.

Another run might produce a smaller mass with substantially greater gold purity.

Which process is better depends on the objective.

This introduces the broader concept of **process optimization**.

Recovery rate, recovery percentage, purity, electrical efficiency, processing time, chemical consumption, and downstream refining requirements all matter.

A process that maximizes only one of these variables may perform poorly overall.

Gold electrolysis therefore provides a particularly clear transition into the final technical chapter of this book. Once recovery has been demonstrated, the question becomes how the system can be controlled to improve its performance.

Last Thoughts

Electrolytic gold recovery demonstrates how electrochemical principles can be used to move gold from an appropriate dissolved form back toward collectible metallic material. Unlike the earlier silver example, the process begins with a prepared gold-bearing solution, making knowledge of solution composition especially important. Low-voltage operation, electrode behavior, mass transport, competing reactions, processing time, collection, washing, and evaluation all influence the result. Most importantly, a dark recovered sediment should not be confused automatically with pure gold. The material must be evaluated within the chemistry and history of the process. With both silver and gold recovery now examined, the final chapter brings these workflows together by focusing on **process control, contamination, purity, and the electrochemical decisions that determine the quality of precious metal recovery and refining**.

Chapter 6: Process Control, Purity, and Refining

Recovering visible gold or silver is only part of a successful electrochemical process. The quality of the final product depends on how effectively the system is controlled from beginning to end. Electrical conditions influence reaction rates, electrode behavior changes during operation, contaminants can move between process streams, and a recovery method that works well for one material may perform poorly with another. This final chapter brings the gold and silver workflows together by examining the variables that determine efficiency, selectivity, and purity. The objective is to move beyond simply making electrolysis work and toward understanding how electrochemical recovery can become a controlled and repeatable refining process.

Controlling Voltage, Current, Time, and Electrode Conditions

Voltage, current, and time interact throughout electrolysis.

They should not be treated as independent settings that can be increased without consequences.

A sufficient potential difference must exist to drive the intended electrochemical reaction. Once that requirement is met, increasing the applied voltage can increase current, but it may also encourage competing reactions.

This was particularly visible in the silver recovery example. Increasing the voltage produced much more vigorous bubbling. Although greater electrical input can accelerate some reactions, visible activity does not necessarily represent more efficient precious metal recovery.

The better objective is to establish an **operating window** in which the desired reaction occurs effectively without unnecessarily promoting competing reactions.

Current provides another important measurement.

Because current represents charge passing through the system, it can be related to the amount of electrochemical transformation occurring. However, the total current displayed by a power supply does not reveal which reactions are consuming that charge.

Some current may contribute to the desired metal reaction.

Some may contribute to gas evolution.

Some may support reactions involving other metals or chemical species.

The proportion of electrical charge contributing to the intended reaction determines how efficiently electricity is being used.

Electrode area provides additional context through current density.

A current of one ampere distributed across a broad electrode produces different surface conditions from the same current concentrated onto a much smaller electrode.

This means that simply recording voltage and current is not always enough when comparing experiments.

Electrode dimensions should also be considered.

DC POWER SUPPLY
CURRENT
VOLTAGE

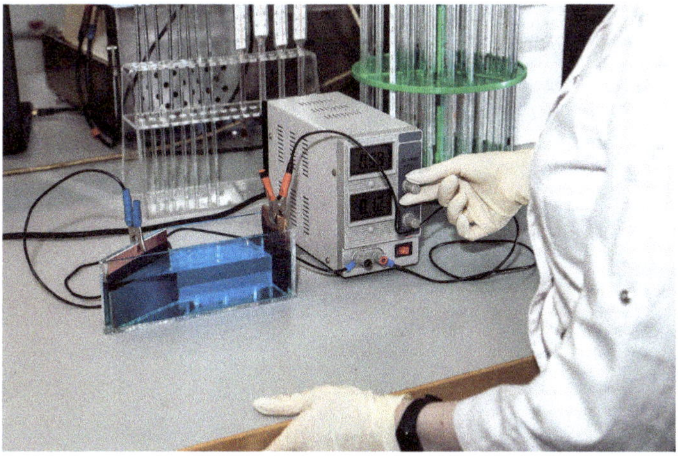

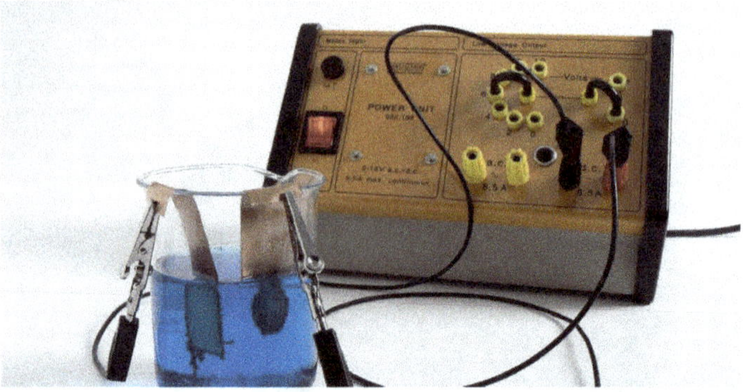

POWER UNIT

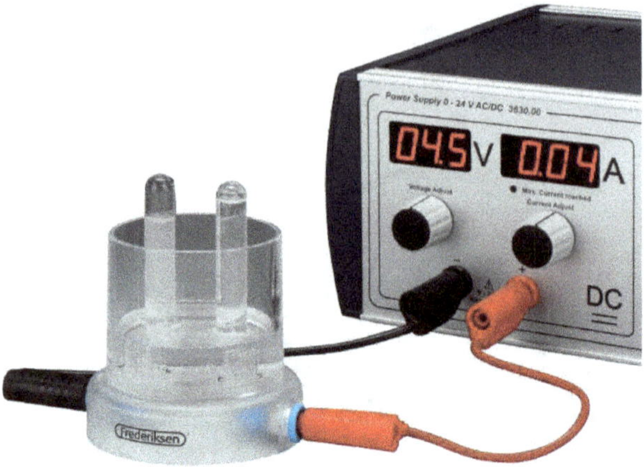

04.5 V
0.04 A
DC
Frederiksen

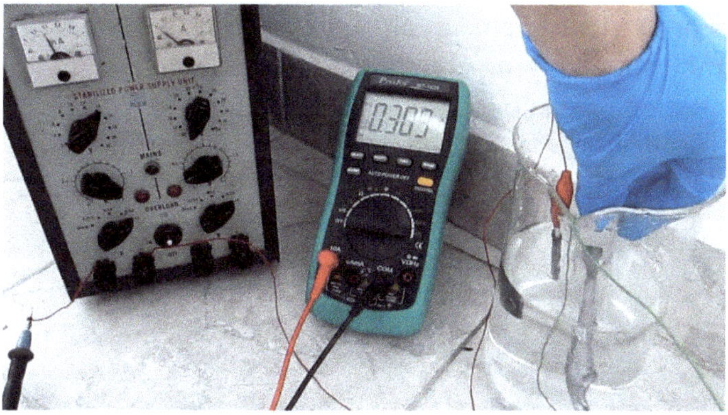

Time presents a similar problem.

A longer electrolysis run does not automatically mean greater useful recovery.

Early in the process, the desired metal species may be readily available and the intended reaction may dominate. As processing continues, the composition of the material or solution can change.

A silver-plated surface, for example, changes as the silver-bearing layer is removed and more base metal becomes exposed.

A gold-bearing solution changes as dissolved gold is progressively removed.

The cell operating at the end of a run is therefore not chemically identical to the cell at the beginning.

This is why observation should accompany electrical measurements.

Current that begins to change significantly can indicate changing electrode surfaces, electrolyte conductivity, concentration, or reaction behavior.

Temperature should also be monitored because increasing temperature can change conductivity, reaction rates, diffusion, and electrode behavior.

The electrodes themselves require attention.

Deposits can alter effective surface area. Surface films can reduce activity. Rough deposits can produce uneven current distribution. Material can detach, exposing new surfaces beneath it.

Electrode spacing may also change if material accumulates substantially.

For controlled work, the operator should therefore think of the electrolytic cell as a changing system rather than a fixed apparatus.

A useful record can include starting material, electrode material, exposed electrode area, voltage, current, temperature, processing time, visible observations, and final recovered mass.

These records make comparison possible.

Instead of saying, "This batch seemed to work better," the chemist or refiner can compare measurable conditions and results.

That is the beginning of process control.

Identifying Contamination and Improving Metal Purity

A precious metal recovery process can appear successful while still producing an impure product.

Contamination can enter at almost every stage.

It may originate in the starting material, electrodes, electrolyte, tools, containers, rinse water, chemical reagents, or subsequent handling.

Base metals are an especially important source.

In silver-plated copper, copper represents a much larger portion of the starting material than silver. If the electrochemical conditions attack the copper excessively, the silver-bearing recovery stream can become increasingly contaminated.

Gold-bearing electronic materials can present an even more complicated mixture.

Copper, nickel, tin, silver, and other metals may accompany gold depending on the original material and the preceding recovery process.

This makes **codeposition** important.

Codeposition occurs when more than one metal is deposited during an electrochemical process. If operating conditions allow another dissolved metal to reduce along with the desired precious metal, the resulting deposit may contain both.

The deposit can still look metallic.

It can still be collected.

It can even melt successfully.

None of those observations establishes purity.

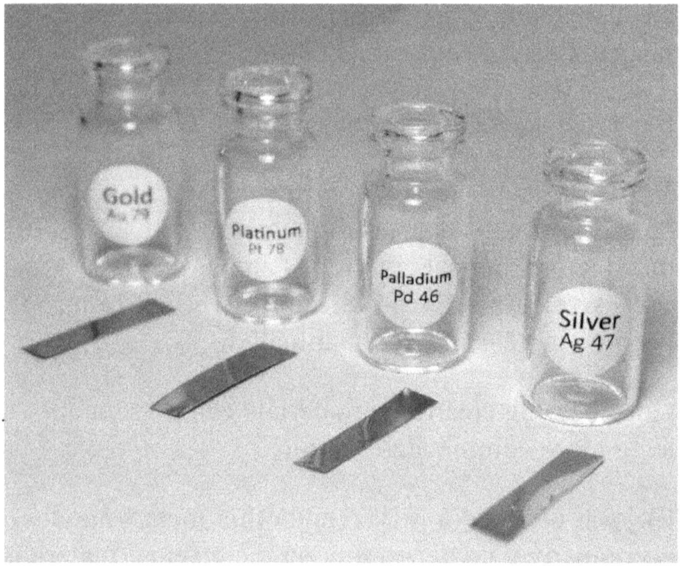

Copper
Cu 29
Gold
Au 79
Silver
Ag 47

Improving purity begins with controlling what enters the system.

Sorting feed materials reduces unnecessary chemical complexity. Keeping different scrap categories separate can prevent one batch from introducing contaminants that would not otherwise be present.

Purity can also improve through selectivity.

If electrical and chemical conditions favor the desired precious metal reaction while limiting base-metal reactions, less contamination enters the recovered product.

Separation between process stages is equally important.

The silver workflow demonstrated this clearly. Electrolysis produced silver-bearing material, but additional processing was necessary because copper accompanied the recovered fraction.

Filtration, precipitation, washing, conversion, and reduction progressively changed the distribution of silver and contaminants.

Each operation created another opportunity for separation.

Washing deserves particular attention because soluble contaminants can remain trapped within fine recovered material. Inadequate washing can leave salts and other residues behind, increasing measured mass and interfering with later processing.

Mechanical losses must also be considered.

Fine gold or silver particles can remain suspended in liquid, cling to vessels, become trapped in filters, or remain attached to electrodes. Aggressive separation of liquids from solids can therefore improve apparent cleanliness while decreasing actual precious metal recovery.

Purity and recovery must be balanced.

This can be represented conceptually as two different questions:

How much of the precious metal was recovered?

and:

How much of the recovered material is actually the precious metal?

These are not equivalent.

Suppose a process produces 10 grams of recovered material containing 90 percent silver.

The actual silver content is approximately:

10 g x 0.90 = 9 g silver

A second process might produce only 9.5 grams of material but at 99 percent silver:

9.5 g x 0.99 = 9.405 g silver

The smaller recovered mass actually contains more silver.

This simple example demonstrates why gross product weight can be misleading.

For high-purity refining, compositional testing becomes increasingly important.

Appropriate analytical techniques can identify contaminants that cannot be reliably detected by sight. Testing also provides feedback that can be used to improve subsequent batches.

If copper repeatedly appears in a silver product, the process can be examined for the stage where copper is entering the precious-metal stream.

If another metal appears in recovered gold, the composition of the original solution and electrochemical conditions can be reconsidered.

Purity analysis therefore does more than grade the final product.

It helps diagnose the recovery process.

Applying Electrochemical Principles to Different Recovery Systems

One of the most important lessons from gold and silver recovery is that an electrochemical procedure should not be transferred blindly from one material to another.

The principles are transferable.

The exact process may not be.

Silver-plated copper presents a particular material arrangement: a valuable surface layer associated with a comparatively large base-metal substrate.

The recovery objective can therefore focus on separating the silver-bearing layer while preserving much of the underlying copper.

Gold recovery from a prepared solution presents a different problem.

The desired metal is already represented by dissolved chemical species. The electrochemical objective becomes reducing those species into collectible material while controlling competing reactions.

Both processes use electrochemistry.

But they do not use electrochemistry in exactly the same way.

This distinction becomes even more important when considering other metals.

Copper electrorefining, for example, can use electrochemical principles to transfer copper from an impure source toward a more purified deposit. Industrial electrowinning processes recover metals from suitable solutions. Electroplating deliberately deposits a thin metallic coating onto another surface.

These applications look different, but they are connected through the same underlying principles of electrode reactions, ion transport, electrical potential, and reaction control.

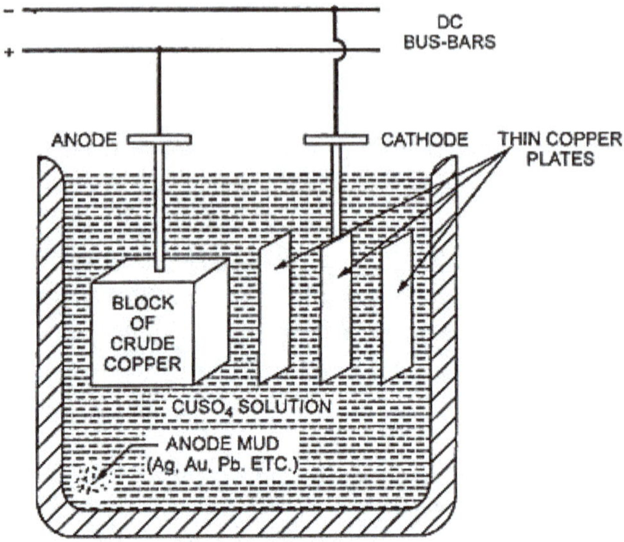

Fig. 8.3. *Electrolytic Refining of Copper*

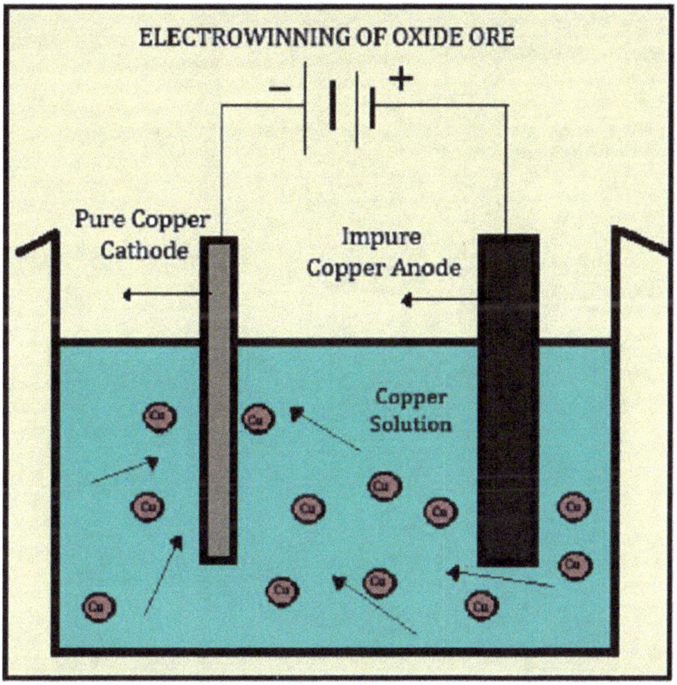

ELECTROWINNING OF OXIDE ORE
Pure Copper
Cathode
Impure
Copper Anode
Copper
Solution

The challenge is determining how those principles apply to a particular material.

Several questions provide a useful framework.

What is the desired metal?

What chemical form is it currently in?

What other metals or chemical species are present?

Which material should undergo oxidation?

Which species should undergo reduction?

Where should the desired metal ultimately appear?

Which competing reactions are possible?

How will the recovered material be separated from the rest of the system?

How will purity be evaluated?

Those questions are more valuable than memorizing a single electrolysis recipe.

They allow a chemist or refiner to analyze a process from first principles.

Consider an unknown gold-bearing recovery stream.

The correct starting question is not, "What voltage should I use?"

The first question is:

What form is the gold in?

Until that is understood, selecting electrical conditions is premature.

The same reasoning applies to silver.

If silver exists as plating over copper, the process objective differs from a solution containing dissolved silver ions. If the silver exists within an alloy, another approach may be required.

Electrochemistry therefore provides a framework rather than a universal procedure.

The more complicated the feed material becomes, the more important chemical characterization and controlled experimentation become.

Small-scale testing can help establish behavior before a larger quantity of material is processed. Variables can be changed systematically rather than simultaneously.

For example, changing voltage while holding electrode area, electrolyte composition, temperature, and material quantity relatively constant makes the effect of voltage easier to interpret.

Changing several variables at once makes it difficult to determine what caused the observed result.

This is basic experimental design, but it has direct refining value.

A refiner who records conditions and changes one major variable at a time can gradually develop a more repeatable process.

A chemistry student can use the same approach to connect theoretical predictions with experimental observations.

That shared methodology captures the central purpose of this book.

Electrochemical theory explains why metal recovery is possible.

Controlled recovery processes provide a physical demonstration of that theory.

When the two are combined, electrolysis becomes more than a laboratory diagram or a refining trick. It becomes a systematic method for understanding and controlling chemical transformations involving metals.

Last Thoughts

Gold and silver recovery demonstrates that successful electrolysis depends less on producing the most dramatic reaction and more on controlling the right reaction. Voltage, current, time, electrode area, temperature, electrolyte composition, starting material, and contamination all influence what the system ultimately produces. Purity must be distinguished from recovered mass, and a visible deposit should never substitute for compositional understanding. Most importantly, the specific procedures used for silver or gold should not be treated as universal recipes. The transferable skill is the ability to examine a material, identify the desired electrochemical transformation, anticipate competing reactions, monitor the process, and evaluate the resulting metal. This combination of **electrochemical reasoning, controlled experimentation, and careful refining practice** is what allows electrolysis to move from a demonstration of chemistry to a useful tool for precious metal recovery.

Last Thoughts: Connecting Electrochemical Theory With Precious Metal Recovery

Electrochemistry becomes especially valuable when its principles can be connected to observable changes in real materials. Throughout this book, gold and silver recovery have provided that connection. Oxidation and reduction explain why metals can change chemical form, electrodes establish locations where those reactions occur, electrolytes allow ionic charge to move through the cell, and electrical energy provides the driving force behind electrolysis. For the chemist, these processes turn electrochemical theory into a physical application. For the precious metal refiner, the same principles explain why a recovery process behaves as it does and why careful control is necessary.

The silver workflow demonstrated how electrolysis can serve as one stage in a larger recovery process. Silver-bearing material can be separated from a copper-rich substrate, but the material collected from the cell should not automatically be considered refined silver. Copper and other contaminants may remain, requiring additional separation, purification, and evaluation before the final metallic product can be properly characterized.

Gold recovery demonstrated a different application. When gold is already present in an appropriate prepared solution, electrochemical reduction can provide a route for moving the gold from dissolved species back toward collectible metallic material. The process depends not simply on applying electricity, but on understanding the solution, the electrode reactions, competing reactions, electrical conditions, and the changing concentration of the desired metal.

These differences reveal one of the most important ideas in precious metal electrochemistry:

There is no single electrolysis process for every gold- or silver-bearing material.

The starting material determines the problem.

Silver plating over copper presents one type of problem. Dissolved gold presents another. An alloy, industrial refining stream, electronic component, or different precious-metal-bearing material may require a substantially different approach.

What remains consistent is the electrochemical reasoning used to understand the system.

A chemist or refiner can begin by asking what form the desired metal is currently in, which other substances are present, which reactions are expected at the electrodes, and where the desired metal should appear after the process.

From there, voltage, current, electrode area, electrolyte chemistry, temperature, and processing time can be understood as variables rather than arbitrary settings.

This is what separates controlled electrochemical processing from trial and error.

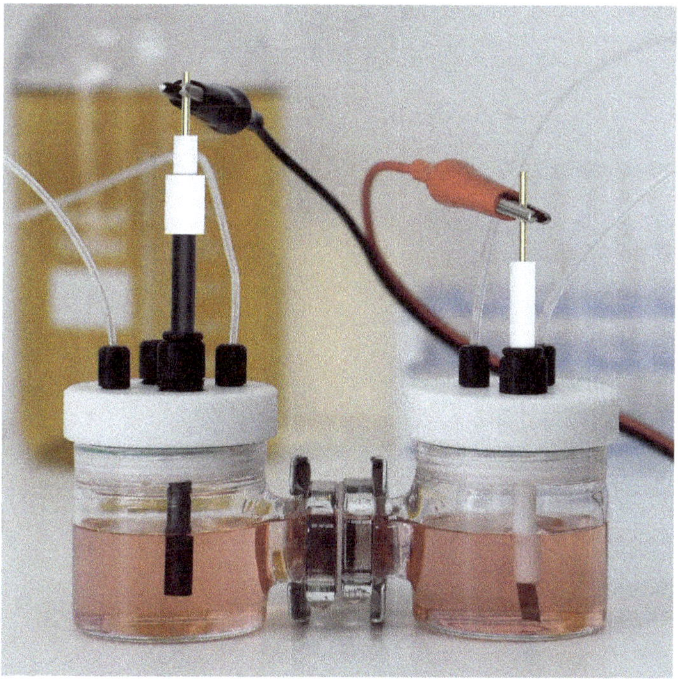

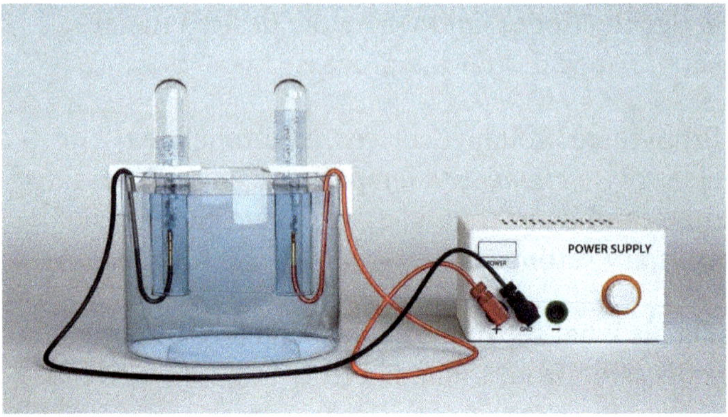
POWER SUPPLY
+
−

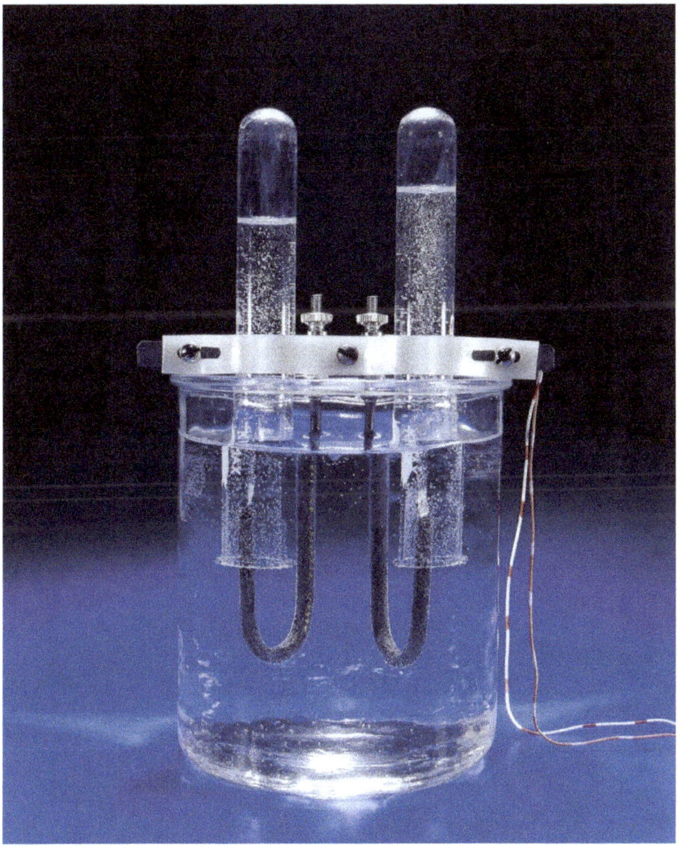

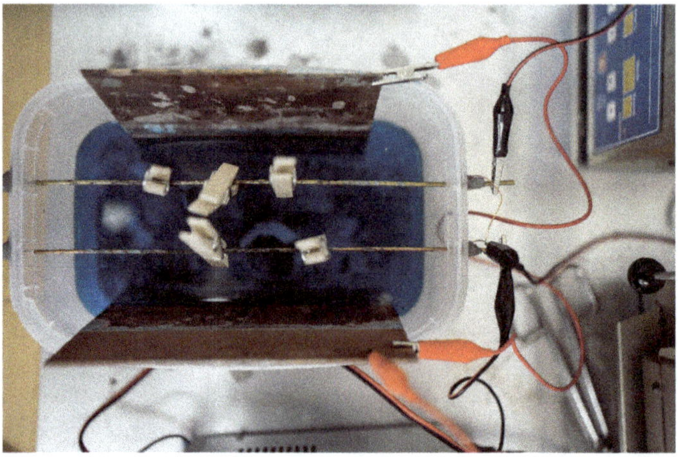

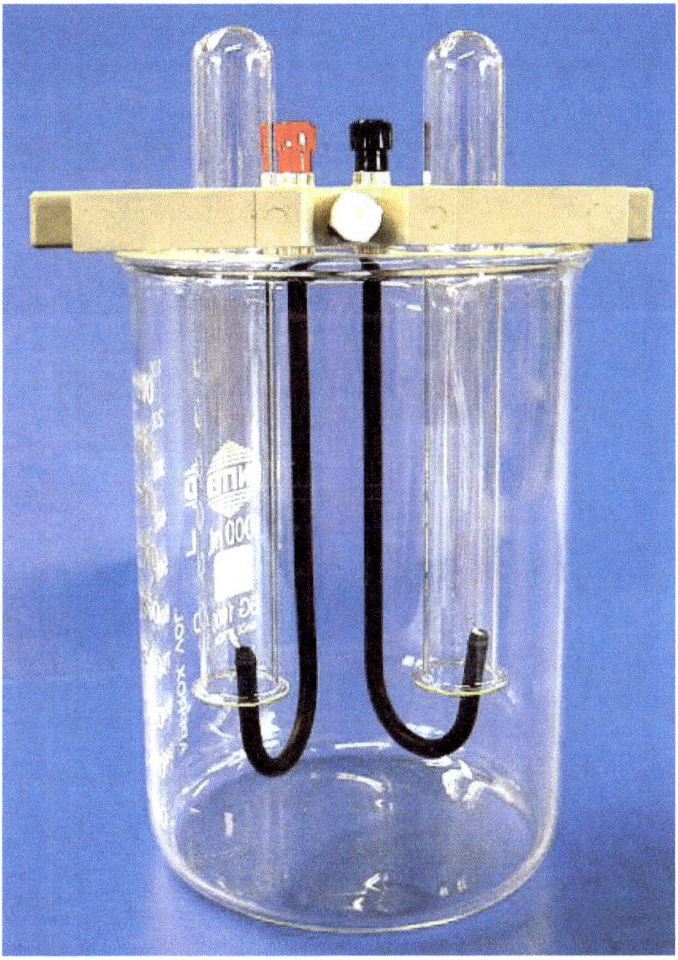

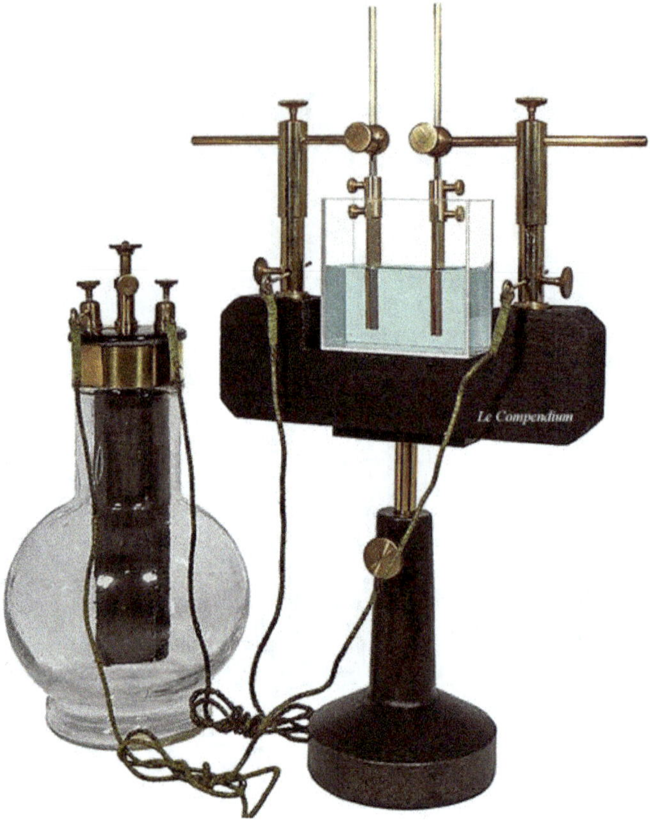

Another important lesson is that **recovery and refining are related but distinct objectives**.

Recovery asks whether the desired metal has been successfully captured or concentrated from the original material.

Refining asks how effectively unwanted materials have been removed from that recovered metal.

A process can perform well at one objective and poorly at the other.

Recovering a large quantity of silver-bearing sediment means little if most of its mass is copper and other contaminants. Likewise, producing a small amount of highly purified gold does not necessarily represent an efficient recovery process if substantial gold remains elsewhere in the system.

The complete process must therefore account for both **yield and purity**.

Mass measurements provide part of that information. Chemical and instrumental analysis can provide another part. Process records connect those measurements to the conditions under which the material was produced.

Together, they allow electrolysis to be evaluated scientifically.

This is particularly valuable for chemistry students and educators. Precious metal recovery demonstrates that concepts such as electrode potential, current density, Faraday's laws, mass transport, precipitation, oxidation states, and reaction selectivity are not isolated textbook topics. They interact within real chemical processes.

For refiners, the same understanding provides a foundation for improving consistency.

Instead of responding to an unsuccessful batch by randomly increasing voltage, changing chemicals, or extending processing time, the refiner can examine the system logically.

What reaction was expected?

What was actually observed?

Where did the precious metal go?

What contaminants were introduced?

What changed during the run?

Which variable is most likely responsible?

Those questions transform recovery into experimentation.

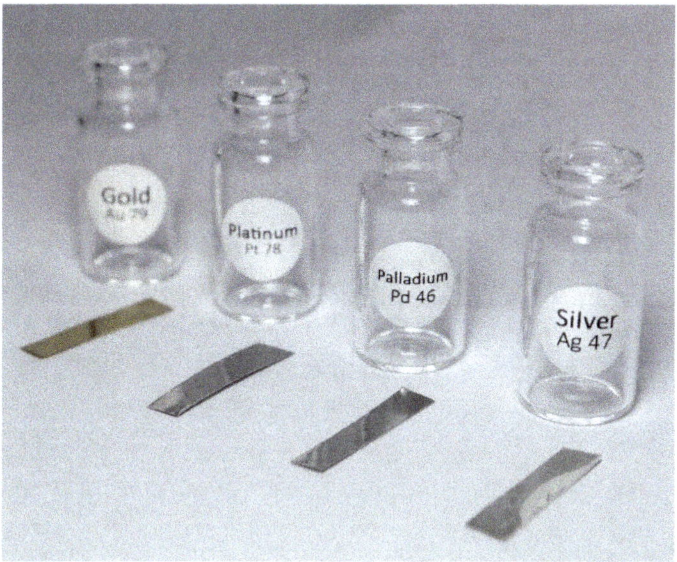
Gold
Au 79
Platinum
Pt 78
Palladium
Pd 46
Silver
Ag 47

Copper
Cu 29
Gold
Au 79
Silver
Ag 47

Electrochemical precious metal recovery also requires respect for the hazards associated with the complete process. Electricity may be only one part of a workflow involving corrosive solutions, reactive chemicals, hazardous gases, dissolved metals, contaminated waste, high temperatures, and molten material. Understanding the chemistry includes understanding that changing an electrolyte, electrode material, concentration, or electrical condition can also change the hazards of the system.

For that reason, technical knowledge should lead to greater control, not greater improvisation.

The strongest approach is systematic: understand the feed material, define the intended transformation, select an appropriate electrochemical system, observe what happens, collect the resulting material carefully, evaluate purity, account for losses, and refine the process based on evidence.

That approach extends well beyond gold and silver.

Electrorefining, electrowinning, electrodeposition, electroplating, corrosion, batteries, industrial metal production, and many other technologies are built around the same fundamental relationship between electrical energy and chemical change.

Gold and silver recovery simply provide an especially tangible way to see those principles at work.

The result is a useful meeting point between two disciplines.

The chemist gains a real-world application through which electrochemical principles can be examined.

The precious metal refiner gains the scientific framework needed to understand and improve recovery processes.

And for both, the central lesson remains the same: electrolysis is most powerful when the person using it understands not merely **that a reaction is occurring, but which reaction is occurring, where it is occurring, why it is occurring, and how that reaction contributes to the final recovery and purity of the metal**.

Additional Resources

Electrochemistry and precious metal recovery bring together chemistry, materials science, metallurgy, and refining. The following resources can help readers continue studying the scientific principles introduced throughout this book and develop a deeper understanding of electrolysis and metal recovery.

Electrochemistry and Electrolysis

Readers who want to strengthen their theoretical understanding should continue exploring topics such as:

- Electrochemical cells and electrolytic cells
- Oxidation and reduction reactions
- Standard electrode potentials
- The electrochemical series
- Faraday's laws of electrolysis
- Current density and current efficiency
- Electrode kinetics
- Mass transport and diffusion

- Overpotential
- Electrolyte conductivity
- Electrodeposition and electrorefining

Chemistry textbooks covering physical chemistry, analytical chemistry, and electrochemistry can provide increasingly mathematical treatments of these subjects.

Precious Metal Chemistry and Metallurgy

Gold and silver recovery becomes easier to understand when electrochemistry is studied alongside the chemistry and metallurgy of the metals themselves. Useful areas for further study include:

- Gold and silver chemical properties
- Noble metals and their reactivity
- Gold and silver compounds
- Metal solubility and precipitation
- Metal complexes and complex ions
- Gold and silver alloys
- Copper and other common base metals

- Precious metal assaying
- Hydrometallurgy
- Extractive metallurgy

Readers interested in refining should pay particular attention to how the chemical form of a metal determines which separation and recovery methods are appropriate.

Laboratory and Analytical Methods

A major step toward advanced refining is learning how to evaluate a process scientifically rather than relying only on visual observations.

Additional subjects worth studying include:

- Laboratory filtration techniques
- Solution preparation and concentration
- Gravimetric analysis
- Titration and quantitative analysis
- Mass balance calculations
- X-ray fluorescence (XRF)
- Atomic absorption spectroscopy

- ICP-based elemental analysis
- Fire assay
- Sampling and contamination control

These techniques can help answer an important question that appears repeatedly in precious metal recovery: not simply whether gold or silver was recovered, but **how much was recovered and how pure it actually is**.

Electrorefining, Electrowinning, and Electrodeposition

Readers interested in industrial applications can continue beyond small-scale precious metal recovery by studying **electrorefining**, **electrowinning**, and **electrodeposition**.

Electrorefining uses electrochemical processes as part of metal purification. Electrowinning recovers metals from suitable solutions using electrical current. Electrodeposition encompasses the controlled deposition of material onto an electrode surface and connects precious metal recovery with fields such as electroplating and materials manufacturing.

Studying industrial applications can provide useful insight into how electrode geometry, electrolyte circulation, current distribution, temperature, solution composition, and process monitoring are controlled at larger scales.

Chemistry Education Resources

Chemistry students and educators can use gold and silver recovery as case studies for exploring:

- Writing and balancing half-reactions
- Identifying oxidation and reduction
- Predicting electrode reactions
- Comparing galvanic and electrolytic cells
- Calculating electrical charge
- Applying Faraday's laws
- Connecting current to theoretical metal deposition
- Comparing theoretical and experimental yield
- Investigating competing electrode reactions
- Evaluating experimental sources of error

Not every precious metal recovery procedure is suitable for classroom performance. Processes involving hazardous chemicals, fumes, contaminated solutions, or high temperatures can instead be examined as **theoretical case studies, demonstrations, calculations, diagrams, or reaction-analysis exercises**.

Safety and Responsible Refining

Anyone continuing into practical electrochemistry or precious metal refining should develop a strong understanding of:

- Chemical compatibility
- Safety Data Sheets (SDS)
- Laboratory ventilation
- Fume-control systems
- Personal protective equipment
- Electrical safety
- Acid and base handling
- Heavy-metal contamination
- Hazardous waste management
- Environmental regulations
- Precious metal refining regulations

Electrochemical knowledge and chemical safety should develop together. Understanding a reaction includes understanding the substances it can produce and the hazards those substances may create.

Recommended Organizations and Educational Sources

Readers looking for deeper technical information can search for educational materials from organizations and institutions such as:

American Chemical Society (ACS)
Chemistry education, electrochemistry, laboratory practices, and professional resources.

The Electrochemical Society (ECS)
Advanced information about electrochemical science, electrodeposition, batteries, corrosion, and related technologies.

National Institute of Standards and Technology (NIST)
Scientific reference data and technical information useful for chemistry and materials research.

United States Environmental Protection Agency (EPA)
Information concerning hazardous waste, electronic waste, metals, and responsible environmental management.

Occupational Safety and Health Administration (OSHA)
Workplace information concerning chemical exposure, laboratory practices, hazardous materials, and personal protection.

University Chemistry and Metallurgy Programs
Open educational materials from universities can provide deeper instruction in electrochemistry, analytical chemistry, hydrometallurgy, extractive metallurgy, and materials science.

Continue Your Research

The processes explored in this book represent only a portion of the much larger field of electrochemical metal processing. Readers who want to advance further can explore academic textbooks, peer-reviewed research, university chemistry resources, metallurgy references, electrochemical engineering materials, and professional refining literature.

When researching a new recovery method, begin with the chemistry rather than the recipe. Determine the form of the metal, identify the other substances present, understand the expected electrode reactions, investigate competing reactions, and determine how the resulting material will be analyzed.

This approach makes continued study more valuable because each new process becomes another application of the same underlying electrochemical principles.

Readers can also find additional books, educational resources, and learning materials by searching "**Roggie Clark Books**" online.

About the Author

Roggie "Raj" Clark is an entrepreneur and a 10-year marketing professional with a proven track record in e-commerce and digital business growth. Having built and managed successful online stores with thousands of sales and consulted for numerous marketing clients, Roggie has mastered the intricacies of online marketing, branding, and consumer behavior. He has also written several popular books on digital marketing, SEO (search engine optimization), and programming (HTML, CSS, and JavaScript).

As a former marketing manager at a fintech company—a role he secured through his own marketing and financial expertise—Roggie gained firsthand experience in financial strategy and wealth-building. This journey inspired him to share insights with amateur investors, helping them save money, make smarter financial decisions, and build long-term wealth. Raj currently lives in Maryland with his sister and niece.

www.ingramcontent.com/pod-product-compliance
Lightning Source LLC
LaVergne TN
LVHW050817061126
851857LV00012B/846